WORLD WITHOUT TREES

WORLD WITHOUT TREES

ROBERT LAMB

PADDINGTON
PRESS LTD

NEW YORK & LONDON

Library of Congress Cataloging in Publication Data
Lamb, Robert, 1949-
    World without trees.

    Bibliography: p.
    Includes index.
    1. Trees. 2. Man-Influence on nature.
3. Forest ecology. 4. Ecology. I. Title.
QK477.L32   582'.16   78-21868
ISBN 0 448 22619 7

Filmset in England by Inforum Ltd., Portsmouth.
Printed and bound in England by R.J. Acford Ltd., Chichester, Sussex.

In The United States
PADDINGTON PRESS
Distributed by
GROSSET & DUNLAP

In The United Kingdom
PADDINGTON PRESS

In Canada
Distributed by
RANDOM HOUSE OF CANADA LTD.

To Tomo

ACKNOWLEDGMENTS:

IT IS A pleasure to thank the following individuals who helped with this work, though their views and mine on the subjects discussed in these pages do not necessarily match:

F.G. Browne, C.J.F. Coombs, Stan Gooch, Brian Greig, Thomas W. Jones, Charles and Sue Manning, Melvin E. McKnight, Clare Oxby, Gordon Robinson, Stephen L. Wood.

Various institutions, too numerous to mention individually, gave help and information: to these also I owe sincere thanks.

Contents

Foreword

THE WORDS YOU are reading now are printed on paper made from trees. This particular paper happens to have been recycled from other paper, but the original paper was made from trees. They were stripped of their foliage and bark and then processed in a complicated operation that extracted from them the cellulose fibers that make up about 40 percent of most commercial types of wood. These fibers were then made into paper. As you turn the pages of this book you will feel the texture of the fibers.

If you put the book down you will be able to mark your place with another piece of paper. If you want to refer back to something in an early chapter when you are midway through the book, you can do so by turning back the pages and scanning them for the information you want. Should you have to go out, and are traveling by bus, or airplane, or train, you will be able to carry the book with you and read it on your journey.

Enjoy yourself. You may be doing something that will be impossible in less than half a century. By then trees may be far too scarce to waste on paper production. Reading will be done on microfilm machines or computer terminals. Books will be collectors' items, and newspapers occasional luxuries. We will get all our news from television screens. Letter writing will be obsolete, except on video readouts linked to telephone or telex networks. Video screens will replace the well-thumbed magazines with which we pass the time in doctors' and dentists' waiting rooms.

In a world without trees, every home will be furnished with metal, concrete, and glass. The ordinary wooden chair or table will have become a rarity. Only the very rich will have access to a park or garden; most of us will no longer be able to enjoy the cooling shade of trees on a warm summer day. Urban oases like Hyde Park and

Central Park will become more akin to barren patches: with their trees gone, they will no longer afford the pleasant illusion of total isolation from the surrounding city.

Children growing up in such a world will never play with a wooden toy — not even a baseball or cricket bat. They will never experience the satisfaction, and frustration, of nailing or sawing a piece of wood. Their Christmas presents will no longer be wrapped in brightly colored wrapping paper. And they will open them under Christmas trees made of aluminum or fiberglass. They will never listen to stories around a roaring campfire, or improvise a fishing rod from a fallen branch. They will never climb a tree, or build a tree-house.

Those children will never see a squirrel or a wild deer; nor will those who live in the country hear woodland birds singing when they wake up early in the morning. For most of the wildlife we take for granted depends on trees for its survival; and as trees disappear it will disappear with them. The robin and the sparrow will be as exotic as the dodo. In the disinfected gloom of museums of natural history, stuffed chipmunks will perch on carefully preserved dead tree trunks behind a glass wall. The children who stare at them may wonder what kind of noise they made, and why they all died. They will see photographs of giant sequoias and redwoods; of the dense tropical rain forests that used to cover vast areas of South America; of New England's spectacular fall display of leaves. On their microfilm readers, they will study the poets who took deep inspiration from nature; but how will they understand it, having never seen a tree, a deer or a forest for themselves? Perhaps they will feel some vague sympathy when they read, in Wordsworth, "That there has passed away a glory from the earth."

For the spiritual nourishment that Wordsworth found in nature will be inaccessible to people living in a world without trees. Nature will have become purely "useful" — a resource for food, economic gain, and energy. Every available bit of rural space will be cultivated for agriculture. The large expanses ruined by the death of the trees will, with great difficulty and expense, be reclaimed so that the human race may be fed. Politics and international relations will be transformed in unpredictable ways. In regions where arable land is scarce, wars may be fought for control of that land.

But all of this, though terrifying, is an optimistic prediction. A world without trees may turn out to be a world in which the human race is pushed to the very edge of annihilation. For green plants, it is well known, convert carbon dioxide — the waste product of breathing and burning — back into reusable free oxygen; without this recycled oxygen we could not live. And unless the atmosphere's surplus carbon dioxide is mopped up by vegetation, the world's climate cannot remain stable. Green plants also play a vital role in the circulation of water, both in the air and in the ground, and in the formation of healthy soil.

Trees, because of their sheer height and mass, are by far the most productive "factories" for these complex operations. Their wholesale death would almost certainly cause a significant reduction in the amount of oxygen available in the air we breathe. At the same time the level of carbon dioxide present in the atmosphere, already much higher than it was a century ago, would increase. Global air temperatures would rise enough to melt the bulk of the polar ice caps; sea levels would go up accordingly; large areas of land, including most major population centers, would be engulfed.

Without trees the pattern of world rainfall would be altered, made irregular; accurate agricultural planning could become impossible. The protection provided by trees to topsoil would be gone, and widespread flooding and landslides — not to mention global famine — could become stark realities. Human life would be a nightmare — if it persisted at all.

You may well be dismissing this as fantasy, as nothing more than conjecture. And, of course, you are right. No one knows exactly what it would be like to live in a world without trees. No one can predict for certain how the death of trees might affect other forms of life.

If you want answers to these questions, you can get them very easily: do not think about the problem, do not try to do anything about it. There is one thing of which you can be absolutely certain: if things go on as they are, some day the sun will rise on a world without trees.

That day is closer than you think.

Richard Ehrlich

1 Wooden Dinosaurs

"The vine is dried up, and the fig-tree languisheth; the pomegranate tree, the palm tree also and the apple tree, even all the trees of the field are withered: because joy is withered away from the sons of men."

JOEL 1:12

A WORLD WITHOUT TREES — the notion has obsessed visionaries in the past. Today it haunts the background of every second science fiction novel or screenplay. Perhaps this means that humanity is aware of the present danger, even if little conscious thought is given to it.

For hardly ever is the question openly asked — exactly what future is there for trees? And how will the world be for human beings if the forests, which made both our evolution and our civilization viable, continue to vanish at today's extraordinary rate?

Over millions of years trees sheltered more than three-quarters of the earth's dry land, their fortunes interrupted only by the severest climatic upheavals. Now, just twenty-five thousand years since the invention of the ax, they cover less than one-third of it. The effect which human progress has had on the world's tree population — mostly during only the last two hundred years — has outdone the worst ice age in thoroughness.

Thanks to the hand of man, the once great forests of Europe are already in tatters. Those of the North American continent are severely depleted. Even the colossal rain forests of South America and tropical Asia are disappearing at a rate that would not have been thought possible a century ago. It is probably already too late to save even these forests in their original form.

Nonetheless, even at this stage, alternative strategies can and must be implemented. Policies, both ecologically sound and economically feasible, can still secure a healthy future for the global environment by placing an irreducible premium on the protection of trees.

One of the tasks of this book is to describe these alternatives. But another is to emphasize how very little time remains. Just as a modern oil tanker requires three miles in which to stop once the order to halt has been given, so even if we shout "stop!" today, the ruin of trees will continue for decades to come. The question is, how many decades do we have left?

This book is really about deforestation. The term is applied not just to the plight of forests as such, but to that of all trees everywhere: in hedgerows, in towns, in gardens, in small woodlands. Moreover, it is applied not only to deliberate exploitation or clearance, but also to the less obviously human influences such as disease and fire. I shall be describing how deforestation comes about, where its effect is leading us, and what we daily lose by it.

The title of this book actually needs help neither from melodrama nor from screeds of technical information to make immediate sense. "Worlds" without trees already exist in plenty. These are the worlds of particular kinds of plants and animals — many beneficial to our own existence — which depend on particular kinds of trees for their livelihoods and which (if not already destroyed) face certain extinction as their habitats are removed wholesale.

In another very literal sense, my title might invoke the already likely prospect of a world so overpeopled as to offer simply no room for trees — or for any plants other than food crops. Statisticians have dated this prospect in earnest at around three hundred and fifty years from now. But how realistic is it to count even on that short breathing space?

A survey of current evidence shows that trees could be so scarce in a mere thirty years' time that they would already be struggling to fulfill their purpose in the global environment. This purpose, as we shall see, greatly depends on their sheer numbers. More to the point, this purpose affects our own well-being in a number of precise ways — through the very air we breathe and the climate we rely on for our food and water. However hard we may find it to visualize a world entirely without trees, we must consider a much more likely possibility — that for all the appalling harm it would bring, even a world without an abundance of trees could fit the bill just as well.

Such a world is not a neurotic image, some kind of desert mirage in reverse. It is a world now unquestionably in the making. It will change our lives in ways we cannot easily accept. But how, and when? The change is already in progress. Can we yet find compelling reasons and manageable ways to reverse it?

Several factors make it hard to answer these questions properly. The most important is the variety of values at issue and consequently the difficulty of attaining agreement from all the interests concerned.

Trees mean different things to different people — to the biologist, to the political economist, to the timber trader, and to the private individual concerned for quite other than professional or business reasons. These are only a few of the more obvious points of view and, I need hardly add, deep divisions exist as much within these interests as between them.

Every separate ethic and attitude relevant to deforestation is powerful and deserves respect in its own way. Even at first sight a universal definition of the problem and a decision on what to do about it seem unlikely to emerge until there are few trees left to discuss. Trees are often loosely spoken of by experts as a "renewable resource" to distinguish them from fossilized forest products like oil, gas or coal. Just how renewable they are we will question.

Environmental issues are notoriously liable to become jammed up by the lobbying of scientific, industrial and business views. For the fact is that most of our problem-solving institutions matured without the benefit of an ecological world view to rival the economic. One well-known version of the position was set forward by Eugene Schumacher in his book *Small Is Beautiful*(29). In Schumacher's judgment, most of the specialist interests that busy themselves with the environment in the developed world and the Third World are attuned to the principle of economic growth. These are therefore bound to be vested interests to some degree — vested, that is, in economics. They have a habit of piously bewailing ecological problems even as they cause them, like the walrus and the carpenter in *Alice Through the Looking Glass*.

Thank heavens that pressure of public opinion can nowadays force environmental issues into the arena of open controversy. Here problem solving often gains in speed and common sense what it loses in precise measurement. An example is the wide public discussion of pollution which has taken place since the last war. This public lobby has produced the remarkable improvement we see in the atmosphere of our cities, and in the condition of at least some of our major rivers.

But then (as far as its fitness as a potent topic for public debate is concerned) the issue of deforestation meets a second stumbling block. For unlike, say, the misuse of pesticides, continuing deforestation is least self-evident where it could most conclusively be debated — in postindustrial North America and Europe.

There are two separate points here that nevertheless strongly support each other. First, deforestation is a historical issue in the West. That is, deforestation has already taken place on a truly vast scale. Second, it is the western, wealthy nations which have most economic and political muscle — the kind of muscle that must be applied (though in an entirely constructive and not in any threatening way) to the countries of the Third World. Only then will they not repeat our mistake — indeed, be forced not to repeat our mistake for the want of food in their mouths.

Little traditional deforestation is currently taking place in the West. On the contrary, what little forest is left is usually preserved with impressive care, even if sometimes along commercial lines that lead nowhere but downhill. So no public outcry arises at home. Most of us fail to observe what is happening to the forests of the Third World — and do not understand how much and how directly it affects all of us.

One very serious form of deforestation does go on in the West in our own landscapes. But it does so discontinuously and obscurely among woods and hedgerows which are scattered and unimposing in their separate extent when compared to a forest. Yet collectively these nonforest trees do add up to a major part of every country's tree population. In fact, they represent tree resources just as important as forests in every palpable respect — but their preservation does not accord with the principles of cost-effectiveness currently in vogue. The problem is not so much that such trees are exploited but that they are not exploitable enough. That is to say, their future is not sufficiently underwritten by market, or by so-called "amenity," values to make their protection worthwhile. Instead, they are regarded as obstacles — to wider highways, new towns and modern farming methods. This despite the fact that these small collections of trees are often composed of exactly the same species that occur in forests and are cared for there, and despite evidence that hedgerows and belts of woodland can do far more good than harm to agriculture.

In Britain alone, over one hundred thousand miles of hedgerow have been purposely obliterated in the past twenty years — the equivalent of several large forests. One authoritative estimate indicates that hedgerows — a clear fifth of the country's present total tree cover — will have disappeared entirely from the British countryside in fifty years' time. Over four hundred species of plants and dozens of kinds of wild animals must follow suit for want of alternative habitats in woods and wildernesses.

The removal of wayside vegetation in parts of the USA in the interest of highway "improvement" has been described in detail in Rachel Carson's *Silent Spring* (2) (see also Chapter 8). The current *remembrement* (boundary rationalization) of French farmlands has also resulted in the uprooting of hedge trees on a vast scale, equivalent to the annual loss of a sizable forest. There is no lack of examples to show that the situation is much the same throughout the temperate northern hemisphere, though it takes different local forms.

All such devastating losses fail, however, to make any vivid or motivating impression on the general public, no matter how far documentary evidence places them beyond doubt.

Meanwhile, in tropical countries where any tradition at all of centralized protest at overt destruction of wildlife forms has yet to develop, the day-by-day clearance of millions of acres of virgin forest proceeds almost unchallenged. It is no exaggeration, as Paul Richards (25) has emphasized, that the tropical rain forests — the largest and richest collections of trees this planet can still boast — will to all intents and purposes have disappeared *in just twenty years' time*.

The Brazilian example springs soonest to mind. During the past hundred years, the colossal rain forests of Amazonia have dwindled by about 40 percent. The construction of giant highways across what remains of the Amazon Basin's unique tree community has now removed all limits on its final exploitation and denaturing (see Chapter 6). Yet even Brazil is not the worst present example in terms of the percentage of forest lost. Many tropical countries have permanently disposed of 70-90 percent or more of their original areas of forest.

Such madcap destruction would presumably not be permitted to happen under modern western eyes in western countries. Yet it is largely because of the industrialized world's insatiable demand for timber (a demand expected to double by 1985) that the devastation continues. When we realize the extent to which we are beginning to rely on the Third World for supplies of cheap wood, we can begin to see why deforestation eludes our fullest attention. We do not want to see the worst and ugliest evidence; we have exported it out of sight.

Some idea of the quantities of timber and land involved can be gained by looking back to the First World War, which by *force majeure* isolated many European countries from the world timber trade. In Britain alone over a seventh of the total forest area had to be removed and consumed for military and domestic purposes.

The present distribution of trees in Europe, perhaps surprisingly, still reflects the maritime or warlike role different nations played up to 1900. Navies are no longer major consumers of timber but general economic pressures on the survival of unprotected trees have never ceased to multiply. Shifts in a nation's political alignment or economic scope can almost be mapped by reference to changes in its forest cover and that of its allies and satellite countries.

Table 1
Woodland as percentage of total land area in 1968.

USA	30.00	Norway	23.8
Great Britain	6.1	Denmark	9.3
Germany	26.8	Netherlands	6.1
France	19.1	Ireland	3.1
Belgium	18.2		

I suggest that economic pressures, remote from their root cause, are threatening also to dispose outright of our own nonforest trees and (in subtler ways) of our forests. If this claim is true, it follows that our technology, which Schumacher plausibly suggests is in a false position at the very best of times, is in no moral position to offer solemn advice to Third World societies on how to take care of their own forests. Until and unless we can create a rationale of protection for all our own trees and abide by it, we cannot ask others to do the same. Still less can we ask others to make economic sacrifices in that cause. Least of all can we ask this when, behind our own backs, we are the very ones who capitalize and encourage the continued destruction of other people's trees — especially of the tropical forests.

It is not just because it is enacted in obscure ways or in exotic places that we are out of touch with the reality of deforestation. For us an extra factor clouds the picture.

Most trees in developed countries are privately owned and their owners are unwilling — or unable — to care for them in life or replace them when they die. No doubt many individuals dedicate themselves to the protection of a few particular groups of trees in beauty spots and along urban boulevards, but the fact remains that most of the woods, hedgerows or avenues we see today in their prime

Table 2
Estimated percentage loss of natural forests since precolonial or preindustrial times. The figures are intended as rough examples; most of them err on the side of caution.

Nigeria	74	China	60
Brazil	40	India	38
Philippines	50	Vietnam	25
Puerto Rico	99	USA	45

are grudged the help they need to perpetuate themselves. They will, on present trends, not be replanted in the future.

This sad and long-established drift must have been difficult enough for the average man to appreciate in past centuries, when local visual evidence or word of mouth was the main source of information. Appreciation is no simpler today, even with the help of statistics and the mass media. For trees are now by their very nature out of step with the normal preoccupations of our society. It has, for example, become unusual for individuals to spend all or most of their lives in one place. And so slow-motion events (like the growth and disappearance of trees) muster few eyewitnesses. Simply because they usually outlive us, trees are still obstinately fixed in our minds as models of timelessness. By the same token, they seem not to need our help. It is true we no longer worship them as our ancestors used to, or as certain civilizations still do. But the underlying myth is still in business. If our own lifespans measured not three score and ten but instead some six score and ten years, the items we should find most conspicuous by their absence in the end would certainly be individual trees.

As things are, and even while we live or enjoy vacations in landscapes still graced by many trees, we may at most feel that familiar surroundings are not quite so wooded, not as three-dimensional, as our early memories paint them. But we tend to dismiss such thoughts in ourselves as pure nostalgia. It is perhaps partly nostalgia. But it is also probably direct observation, inevitably less than keen. The science fiction vision of a treeless landscape builds its box-office, I think, on precisely such ambiguous misgivings.

In future the desolated landscape will speak for itself in no uncertain or ambiguous way. For the present, however, its fate still remains partly hidden in the time-shadow between tree and human generations.

If we could see through that shadow, prospects for the tree would — in my opinion — automatically improve. Solely for the sake of their sentimental value, the public's collective responsibility for non-forest trees would increase without any further protective custody in the form of public ownership or legislation. For I believe that trees hold for all humans a rock-hard fascination that no amount of technological or popular trivialization can ever completely erode. I also feel that, far from being irrelevant or fanciful, such feelings must be regarded as just as real a resource for securing our environment's future health as the best-endowed technical know-how — no more and no less.

I have in mind, by way of evidence, exceptional examples of developed countries which have suffered the spectacular loss through exploitation of true natural forest in recent years. There are also areas where so few trees exist in any case that deforestation, in any guise whatever, cannot fail to be conspicuous. In New Zealand, for example, which has lost over 75 percent of its very large forests during the past hundred years, and Australia, which is now only 2 percent but in 1900 was 4 percent forested, first-hand, active interest in the well-being of trees appears to be spreading throughout society. The protection of wilderness is becoming a daily concern of trade unions, business interests and politicians alike — a sign that market and nonmarket values are switching places in response to informed public opinion. A successful public lobby in the early 1970s to prevent the despoilation by Rio Tinto Zinc of forested wilderness in Australia is a case in point.

When people in democracies can see what is happening, they show how deeply they care about trees and that they are prepared to do something effective to protect them on a scale that can be believed in. Must we, though, wait until the loss of our homeland trees becomes spectacular enough to offend our sense of landscape? Or until the side effects of worldwide deforestation begin to reach into our lives in more practical and horrifying ways? Is this what must happen before we embark on our own conscious reappraisal of what trees mean to us?

I for one hope not. Accordingly, I use this book to dramatize the plight of our localized, nonforest trees by placing it in perspective with world deforestation — and vice versa. In so doing I shall draw out the full implications which the situation holds, not only for our own future, but also for that of our children.

My task is made easier — though in another sense more cheerless — by the present occurrence throughout the western world of epidemics of Dutch elm disease. It is a problem which seems at first sight to have little in common with the frenetic axwork now in progress among the rain forests of the Third World. In reality, however, both these aberrations result from the same combination of human improvidence with outright error, and both raise almost exactly the same ethical and biological issues. They differ only in direct cause and extent.

In Kurt Vonnegut's *Breakfast of Champions*, an embittered truck driver denounces God as a lousy conservationist and cites Dutch elm disease to prove his point. He could have chosen better examples, for Dutch elm disease is no act of God. We can say with certainty that if there were no modern timber industry, there would be no elm disease epidemic across the northern hemisphere today.

I shall clarify these statements later. I shall, in fact, use the many aspects of Dutch elm disease (DED) as openings for the airing of questions posed by deforestation as a whole.

One justification for my approach is, as stated, that DED is obvious and topical. Most North Americans or Europeans need not go very far to see for themselves what the problem looks like, and to judge for themselves the difficulties that have so far stood in the way of a solution. Perhaps they may work out for themselves ways in which they personally can influence the course of events and then take an individual initiative against deforestation which other members of society can imitate.

Before discussing DED, we need to be quite clear that elms are in no way the only temperate trees prone to epidemics of parasitic disease. They just happen to be one of the worst affected at the moment. At present about one in eight of the world's elms dies each year as a result of DED. This statistic means in real terms that by 1984 elms will probably be such rarities that they can no longer play any effective part in ecosystems, in landscapes, in agriculture or commerce.

By that time it is quite likely that other trees (notably oaks, beeches and chestnuts) will then be in much the same predicament that elms are in today. Many tree species have already begun to succumb in quantity to diseases as complex and potentially just as destructive as DED.

The stark reality is that all the trees (both hardwood and softwood) native to cool temperate zones of the world seem to be under unprecedented pressure from disease of one sort or another. Disease is taking on almost as destructive a role here as commercial exploitation is in the Third World. It is doing so for reasons that are partly natural and partly the result of human interference.

The unnatural factor in the equation derives basically from the hectic rate at which human commerce and mobility have increased during the nineteenth and twentieth centuries. An important consequence of this ceaseless trend is that people have contrived more and more to remove certain plants and animals from their normal surroundings and transport them across natural barriers into places where they do not belong. Natural stresses oppose the survival of any organism in any locality. Against the stresses present in a quite unfamiliar place, however, the transported organism is unlikely to have natural defenses. Most organisms only possess defenses appropriate to the precise conditions in which they have evolved over geological time. But some introduced organisms are all too successful and monopolize resources at the expense of native forms.

Every living thing in its natural habitat forms part of an ecosystem. An ecosystem is a localized network of plants and animals interacting with the climate, and with each other, in recognizable and continual cycles governed mainly by the character and availability of the most basic of all food resources — the dominant vegetation. It is the vegetation which calls the tune as far as all animals are concerned. The vegetation is in turn heavily dependent on weather and soil conditions.

For example, an abnormally cold summer in a grassland region will result in a shortage of the grass seeds of various kinds which form a basic item in the diet of some small rodents. By reason of the shortage of food, the rodents in question will breed in smaller numbers. In turn the populations of hawks, owls, foxes and so on, which feed on the rodents, must also be reduced.

Still this example does not show us the full complexity of what is involved. Animals in their turn affect the soil and the supply of vegetation. Without the action of worms, insects, millipedes and so on, the soil loses a major part of its ability to produce vegetation. These soil organisms feed on the excreta of animals that live above ground, and on fragments of dead vegetation, including those dropped or partly dismantled by plant-eating animals above ground. The ways in which energy is cycled in one form or another through the living world are endless but theoretically measurable.

Such intricate patterns of interdependence define what we call ecosystems. The concept of the ecosystem can be applied on any scale from the ecosphere (the entire biological energy-exchange system of a whole planet) downward.

Because they are somewhat isolated from one another, major continents represent distinct ecosystems on a very large scale, which are then internally subdivided into smaller and still smaller systems. The smallest system of all is the hard-won ecological niche of individuals of each unique species of organism — including the most microscopic of parasites.

The integrity of ecosystems can be breached by many unforseeable circumstances — as when a freak storm carries an African locust or butterfly swarm to islands in the Atlantic, or when movements of the earth's crust create oceans between areas once linked together, or when rats from an explorer's ship run ashore into a new land.

Human agriculture has a far more deliberate effect and works fundamental changes across a whole range of ecosystems, when it supplies otherwise barely competitive plants and animals artificially and continuously with conditions they now seem to dominate naturally. These operations are more or less successful in the short term and the narrow sense, or we would probably not be here to tell the tale — hungry or not. In fact, two out of three people in today's world are starved or half-starved most of the time.

Generally speaking, however, when an organism is taken from its proper habitat and transferred to an alien ecosystem, the result is likely to be trouble and enforced change for the organism itself or for the organism's new neighbors — and in the not so long run, trouble for us.

This statement holds true when, say, bacteria leave a July oyster where they do no harm and enter human intestines, where they can do plenty. It was equally true when the European rabbit was introduced into Australia during the nineteenth century, and subsequently became the farmer's worst enemy there. It was again true when certain South American water plants were introduced into the Congo and Zambesi rivers in the 1950s for ornamental purposes, and soon made major river systems unnavigable and destroyed the productivity of lake fisheries. It was yet again true when the fungus that causes DED was twice shipped to and fro across the Atlantic, a stowaway in insanitary commercial timber, and on each occasion started or reactivated epidemics in America and Europe.

It might be pointed out that the importation of rabbits into Australia or waterweeds into Africa seem to be cases of unwise deliberate introduction, while the examples of food-poisoning bacteria and of DED seem on the other hand to be cases of accidental introduction. As such, those were hardly anybody's fault. But today common knowledge of the biological dangers of importing timber pests has broadened so much that we can no longer classify or excuse current introductions of tree parasites as accidents, any more than we can call it an accident when a restaurateur knowingly serves a customer with an out-of-season oyster.

Legislation does exist in Europe and North America to impose quarantine checks on imported timber. But such checks are difficult to enforce and are often ignored with impunity. Greater self-regulation on the part of the timber industry seems to be the only solution.

There is a long queue of tree diseases (led by oak wilt and chestnut blight) which threaten to become major cosmopolitan epidemics in the wake of DED. At present, most of these are confined to one side or the other of the trade oceans. But oceans are ponds to modern shipping and air freight and it seems only a matter of time before these diseases also spread to neighboring countries and to distant continents unless quarantine restrictions are improved and respected.

Even if this were the case, it should not be forgotten that timber is an exceptionally awkward material to clean and containerize. No matter how conscientious timber importers became there would always remain a risk from cases of genuinely accidental introduction. Sheer chance, as history shows, can work miracles of destructiveness.

When prevention fails, what hope can be held out for cure? To judge from the example of elm disease, very little. If we cannot save the elm, it seems unlikely that we shall manage to save the oak, the beech, the chestnut, or even sought-after commercial softwoods as and when such trees fall prey to epidemic doses of parasitic diseases that belong to the same large category of ailments as DED.

Nor can we say with certainty that DED itself will eventually disappear for lack of elms. There are abundant precedents for fungal parasites and bark insects adapting successfully to new tree hosts and intermediaries when the initial host is gone. So far, DED has only been known to affect one kind of tree other than the elm — the closely related and uncommon zelkova. But there is, strictly speaking, no overwhelming reason why it should not show some biochemical versatility and attack, say, apple trees, possibly through the agency of a different bark beetle.

There is, furthermore, growing evidence that viruses, as well as insects and fungi, are implicated in the process which turns diseases like DED into epidemics. The bizarre biology of viruses includes a capacity for mutation that is often completely unpredictable. Lewis Thomas (34) has speculated that viruses may be: ". . . mobile genes . . . a mechanism for keeping new, mutant forms of DNA in the widest circulation among us." This idea we shall examine later for its possible relevance to the evolutionary history of trees. In the context of this chapter, it is enough to say that all or any of the parasitic agents that add up to DED could play a part as yet unguessed in future plant diseases.

This circumstance, in itself, constitutes an argument in favor of our trying harder to save the elm. If we can win this battle, we may overcome or perhaps even avoid later shocks.

If a new approach to that task is not soon found, however, if we finally admit defeat on grounds of false economy, the elm may subtly leave behind it problems that could make DED seem in retrospect insignificant. It could profit us to recall that it took only one wooden horse to completely surprise all the fabled and stabled livestock of Troy.

So much for direct and indirect human influences on the health and fate of trees like the elm. What nonhuman factors might help to account for the current prevalence of tree diseases in the temperate world?

It is well known that trees native to cool climates are always more vulnerable to pest outbreaks than trees in natural forests on or near the equator. One of the reasons proposed is the lack of species diversity among temperate trees. Tropical rain forests contain hundreds of different types of large tree — 375 species of tall hardwood were counted in one area of Malaysian forest the size of ten football fields. In a similar area of temperate forest, one would have a hard time finding more than a dozen or so species. There are other reasons also to which we shall return.

What I want to touch on at this point are the implications not so much of the intricate ecological mechanisms at work in forests at present, but in the most general way of the history of trees across the last five hundred million years.

A glance at the geological chart, Figure 1, gives the orthodox view of such trends down through the ages. The tree-fern forests of the Carboniferous, source of the bulk of the fossil fuels (coal, oil, gas) we use today, ceased to dominate the earth's vegetation roughly two hundred and fifty million years ago. They were supplanted by primi-

Figure 1

Chart of geological ages, simplified. Seed-bearing plants are classified as gymnosperms (having naked seeds, as pines) or angiosperms (with seeds encased in flowering structures, as elms). Cycads (now almost extinct) are intermediate in form between pines and ferns.

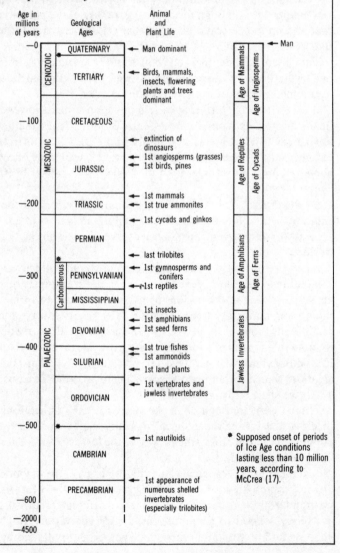

tive conifers, pines, cycads (trees which had features in common with modern pines and flowering plants), and then by flowering plants, that is, by grasses and others, including the flowering trees we rather inaccurately call "hardwood," "broadleaf" and "deciduous," in order to distinguish them from conifers.

Successive heydays of different types of vegetation are characterized in the chart as the "age of ferns," the "age of cycads" and so on. These can be seen to coincide with comparably important reshuffles among the earth's animal life — the ages of amphibians, reptiles and mammals being the conventional landmarks.

It goes almost without saying that the connections here are causal, for trees do play such a stable and important part in translating environmental habitats for smaller, shorter-lived, mobile creatures that they *are* the environment, in a manner of speaking, for the purposes of the evolution of most land animals. Climatic and geological upheavals, which found their most extreme expression in ice ages and in independent movements of the earth's major land masses, were of course major turning-points of evolution, but the biological changes they worked on land took place largely through their effect on trees.

Fairly recent research into the history of climate suggests that ice ages have occurred during the last five hundred million years in three batches, marked by asterisks on Figure 1, each batch lasting only a few million years and succeeded by about two hundred and fifty million years of relative climatic calm(7).

It can be seen that the assumed chronology of these ice age epochs also coincides to some extent with changes in the character of the world's forests and fauna, and that the present day marks the recent end of one of the three flurries of ice movements.

Many inconsistencies make it difficult to draw complete or absolutely precise causal connections between such events and the extinction or prospering of different plants and animals. Darwinism and geology can go some way toward explaining these inconsistencies, but the usual conclusion is that one can never hope to apply full ecological scrutiny to what is an incomplete fossil record.

That incomplete record, though, gives many fascinating clues to the effects of climate, planetary motion and the contortions of the earth's crust upon evolutionary history. Yet it gives us very little inkling of the specific influence *disease* might have had on the development of trees up to the present, and therefore also on the destiny of most other terrestrial forms of life.

It seems highly likely, however, that trees — both on and off the equator — must have been even more prone in past ages to pandemic diseases than they are today. For until the present continents finally separated from one another, some fifty to a hundred million years ago, there were fewer geographic barriers to the spread of parasites. Moreover, early forests were not divided by clearance as they have been since man evolved, and were probably also less diverse than at present.

Fungal or viral parasites do not, as a rule, leave clear evidence of their activities in the form of fossils. Yet there is no doubt that they have been present in one form or another on the earth for much longer even than trees. Is it possible that this factor — think of the effects of fungal parasites — could account for some of the observed inconsistencies we find in the evolutionary picture?

What I am trying to suggest is that if aberrant (e.g., ice age) climates have altered the course of evolution through their effects on the evolution of trees, they may also have done so — as often as not — by way of wholesale disease epidemics long after the climate itself had stabilized. Such epidemics may have come about among trees whose defense mechanisms were acclimatized to bygone shocks and which were therefore outmoded in relation to the activities of quick-breeding, quick-mutating parasites. My implication is that the present wave of disease affecting North American and European trees may be, in part, a delayed response to the events of past ages — specifically of recent ice ages.

A further implication is that the current deforestation of the planet by man may not be a flying in the face of nature, but, on the contrary, a speeding up of a natural process, which would be happening even if humans did not exist — but which would happen more slowly.

This does not mean that we cannot still call the shots — it is a question only of fully understanding the shots we call. Even so, we are making things mighty hard for ourselves.

The tropical rain forests, secure from the worse effects of climate and disease, have hitherto always been around to recolonize land laid bare by extreme global climate changes, once more normal conditions are reestablished. But their age-old ability to act as rallying points for forest life forms is very much at risk from the exploitation they themselves are suffering today at the hands of man.

The logical conclusion of the two trends I have described is the possibility that trees — unless we are prepared to reverse our present policies and protect them — stand to become the dinosaurs of the present era.

Most of the points raised so far are not only open to critical debate, but are also liable to meet with the cynic's ultimate response: "So what?" The question implied is: do we need trees at all? Or is the advantage they bestow on us or the planet so marginal that such advantage cannot justify the expense involved in protecting or ceasing to exploit trees?

The simple answer is that nobody knows for sure. But in this book I shall undertake a thorough questioning of the role trees play in the making of the weather and the atmosphere, in marshaling water resources, in facilitating farming, in medicine, in commerce, for we dare not be wrong. We dare not let trees go, and only then realize that we have made a fatal and irrevocable mistake.

The heart of the matter is, in my own view, really an ethical as much as a technical or economic set of choices. The scientist Eugene Rabinowitch has stated that:

> The only animals whose disappearance may threaten the biological viability of man on earth are the bacteria normally inhabiting our bodies. For the rest there is no convincing proof that mankind could not survive even as the only animal species on earth. If economical ways could be developed for synthesizing food from inorganic raw materials — which is likely to happen sooner or later — man may even be able to become independent of plants, on which he now depends as sources of food . . . .

Between the world Rabinowitch hints at and the world we know lies an inscrutable but possibly very short span of time. A century? A hundred centuries? Whether it will come and just how long it takes depends partly on choices we make here and now. Even supposing it is possible, do we actually want an azoic world for man only, where conditions are everywhere the same as those we see today in any large industrial city?

If we do, if we can happily accept such a prospect, only then can we tolerate our own present inaction. Only then can we afford to ignore the possibility that could represent the most irrevocable step of all toward it — the possibility of a barren world without trees.

2 Trees and the Quantity of Life

"Woe unto them that join house to house, that lay field to field, till there be no place, that they may be placed alone in the midst of the earth."

ISAIAH 5:8.

IN TERMS OF sheer weight, over seven-eighths of the living matter now occupying space on our planet consists of plants. A relatively tiny remainder of animal life, including human beings, obtains all its nourishment directly or indirectly from the, in many ways, self-sufficient plant kingdom. All these organisms weighed together make up the global biomass.

The relationship between these many forms of life and the atmosphere or water in which they live is extremely complex and can be extremely delicate. The alteration or removal of one tiny or apparently trivial item from the total equation can have incalculable effects on other organisms all along the line.

Animals draw their nourishment from plants, either directly by browsing or indirectly by feeding on other animals which eat plants. Only green plants themselves live directly on inanimate matter — water, sunlight and air. And yet, for the maximum development of their potential, green plants also rely on animals and lesser plants, and especially on the organisms which digest and recycle dead organic matter (both plant and animal) into a fresh supply of nutrients for the plant kingdom. Broadly speaking, these organisms are the bankers of the total system, always ready to supply fresh capital for new ventures.

Plants, we said, represent about seven-eighths (or 85 percent) of the world's biomass. Trees, in turn, represent 90 percent of the plant biomass. These figures alone give us some idea of the key role played by trees in every living transaction that takes place on our planet. How can we, for a second, contemplate their removal?

If trees at present account for 90 percent of the earth's plant tonnage, the tropical forests in turn constitute a very large part of this whole. Such statistics as these seem to give little reason at first glance for pessimism about the extent of our forests and woodland. But we have to consider them in perspective. The present-day predominance of trees in the world's vegetation is only a shadow of what it was before agriculture became a part of the human way of life and a ghost of what it was before the Industrial Revolution. The world's tree cover and the biomass in general have decreased as fast as the human population has increased — and the greatest toll has therefore been taken during the past century, during humanity's technological (and so population) explosion.

A hundred years ago the earth's dry land was (according to some experts) made up of 42 percent forest, 34 percent desert, and 24 percent grass and agricultural land. Today the relative proportions of these land types one to another is reckoned at 33 percent, 40 percent, and 27 percent respectively. But the word "forest" nowadays includes some extremely thinly wooded landscapes, among them areas of marginal land that could, under pressure from a mere year or two of drought or intensified man modification, give way directly and irretrievably to desert.

So while tree cover up to the Stone Age was typically massed in near-continuous forests, except in the world's arid and polar regions, it nowadays takes far more scattered and tenuous forms. The residual woods that form a rather one-sided mosaic with modern temperate farmlands, the sprawling "taiga" conifer stands that fringe the Arctic Circle, the scrubby chaparral regions of the American Southwest — all these are theoretically components of the world's "forest." The area they collectively span on the map actually exceeds that of the colossal forestation of the humid tropics. The density of trees per acre is a very different matter, and every day this factor becomes less and less homogeneous worldwide.

Forest and so-called forest (whatever form that takes) is nowadays generally reckoned to shade about ten thousand million acres of the globe's surface. However, as usual, it is impossible to claim exact figures on this score. There is no truly universal and reliable record of land use, and forest areas undergo change and reduction with every passing day. Estimates published by international agencies concerned with land use are tentative — to say the least — drawing their data from out-of-date maps or questionnaire returns which often involve guesswork.

The interpretation now in progress of satellite photographs, sophisticated enough to distinguish between different types of vegetation on the earth's surface, should produce the ultimate statement of account within the next ten years. By that time, earthbound estimates suggest, the tree cover will already have diminished by an overall 20 percent, and locally, especially in rain forests, by more than half.

About 2 percent of the remaining total lost per year is the most authoritative figure currently given for the rate of deforestation worldwide(14). While there are certainly a few technologists who still deny this figure is credible, they are outnumbered by those who consider it reasonable.

THE DAILY TRAGEDY OF DEFORESTATION
Even in the best-regulated commercial plantations and national parks, it is generally true to say that many more trees are destroyed yearly than are planted yearly, even disregarding "infant mortality" and the effects of routine thinning and cropping. Drought, disease and fire are the main causes of loss, but these "natural" phenomena obtain considerable encouragement from human activities. Diseases, for example, are spread well beyond their normal range by careless trafficking in diseased lumber between neighborhoods, states and continents. Atmospheric pollution and the canceling-out of natural enemies of tree pests, through both the on- and off-target spread of agricultural pesticides, also conspire to weaken the natural resistance to disease of many trees.

The campfires and cigarette butts of vacationers, refugees from overcrowded cities, are the prime sources of forest fires. But there are also less noticeable human depradations. The tapping of lakes, rivers and underground reserves of water for our urban and service reservoirs impairs the ability of millions of trees to survive drought conditions. And drought itself in a sense is an upshot of deforestation in general — since there is no doubt that forests, or the lack of them, play a large part in determining the incidence of rainfall and groundwater. The routine cropping of timber in commercial plantations theoretically strikes an even balance between felling and replanting, but, in practice, harvesting almost inevitably exceeds input in the long run. Certain large timber-marketing corporations and nationalized forestry agencies make a boast of planting two, four or more trees for every tree trunk felled. If this were true, then in some developed countries, where this good intention has supposedly been enshrined in practice or in law for a century or more, every last nook of available land should now by rights be covered in trees twice or thrice over! The stark fact is that, in almost every western country where authentic statistics are made public, these now show a steady decline in the managed woodland mass. Sometimes of course the symptoms of decline are less overt. Witness the fact, for example, that the trunk diameter of commercially grown trees at market time has everywhere decreased by an average quarter or more during the past thirty to forty years. In other words, though the numerical decline of managed trees may in some countries (including the USA) seem negligible on paper, the amount of renewed forest permitted to grow to maturity, or to the peak of its ability to affect the environment, shows an unequivocal loss from year to year.

In unmanaged woodlands, where conservation legislation and business standards hold less sway, the loss is even more dramatic, though admittedly not precisely measurable. Hedgerows, local wildernesses and private woodlands that do not fall within the category of managed forests, form in most developed countries the bulk of the overall tree cover. Road building, urban sprawl, the enlargement of agricultural fields, natural death not compensated by replanting — these and hosts of other, less obvious, factors are, it stands to reason, more successful than is management at "mislaying" trees.

The death of trees in the developed countries was the very thing which from knowledge of fire and the wheel onward made civilization possible. The discovery of agriculture in Europe between five and ten thousand years ago (following the Middle East's much earlier lead) set in motion a chain reaction of forest destruction. It continues to this day, although it has taken a multitude of different forms at different times. The smelting process that gave the Iron Age its *raison d'être* required the use of charcoal (cooked wood) in industrial amounts. Iron tools then of course made agriculture and deforestation still easier. The Industrial Revolution, though it obtained its driving force from fossil rather than from living forests, accelerated population growth and made further destruction of the treescape inevitable, the more so because it was now mechanically aided.

The exhaustion of forest resources had, long before this, become one of the most telling motivations for the European colonization of other lands. Even in the thirteenth century, legislation designed to protect forests from indiscriminate felling and from grazing by domestic cattle existed in twenty-eight states of Europe. Deforestation was already a recognized problem, but mastery of navigation and shipbuilding technology soon made it possible to look for solutions overseas rather than at home. In Columbus's time, an estimated eight hundred million acres of virgin forest stood in what is now the USA mainland. The first European settlers were astonished by such richness and the bonus of agricultural land the situation betokened. They did not let the contrast remain astonishing for long. By 1930, the forests of the USA occupied only one hundred million acres; today the figure is closer to fifty-five million acres, despite conspicuous attempts at reforestation since the war years. The systematic removal of the North American tree cover before 1900 bears exact comparison to what is happening in tropical forests today, except that there it is proceeding at a twentieth-century rate.

The developed world is not, and never has been, secure from deforestation. Nor has it dropped the habit of exporting the problem, for it is in a quite definite sense responsible for deforestation in the tropical Third World — where the most spectacular losses of trees are today sustained. Commercial exploitation of tropical forests partakes to a large extent of western technology and funding. Yet the most influential factor in tropical deforestation is neither technological nor truly commercial in principle.

It surprises many people to learn that most of the trees annually trimmed off the face of the earth for human purposes are required not for paper production, not for construction timber, but for firewood or simply for the sake of the arable soil their roots embrace. These last uses somewhat amazingly account for about 46 percent of the current world deforestation total.

In areas still heavily forested — in the Amazon Basin, for instance, or in parts of Central Africa — the chief pressure on the survival of forests takes the form of shifting cultivation — the slashing and burning of woodland tracts to clear the land for temporary, exploitive agriculture. Temporary because soil cleared of its virgin forest cover remains fertile for only a short while. The subsistence farmer cannot, by definition, afford to buy fertilizers, and must soon cut his way further into the forest if he is to grow enough food to nourish his household. He leaves behind him a pauper terrain which may take up to two and a half centuries to regain its original appearance and fertility, if by some chance it is left undisturbed by subsequent farming or human settlement.

In areas such as northern West Africa, where forest has long ago given way under similar pressures to a mixture of sparse grassland and scattered drought-resistant trees, the main pressures are cropping for firewood, destruction of saplings by overconcentrated herds of domestic animals, and the prevalence of bush fires. The inhabitants of a single city, Kano, in the northern Nigerian savanna, crop seventy-five thousand tons of firewood every year from the surrounding woodland. If this kind of figure applies on the same scale to towns throughout northern Africa, it is no wonder that desert is being formed there at a quicker rate than anywhere else in the world. Besides tempering the climate, savanna trees act as rallying points for the regrowth of overgrazed or razed grassland. Their removal inevitably encourages desert encroachment. It has been estimated that, until only five thousand years ago, the western and central areas of the Sahara were covered in woodland, of the type which now exists in Mediterranean countries and is characterized by the word *maquis*. It is likely that grazing, farming and woodcutting were main causes of its disappearance from the area we know as the Sahara Desert.

While the industrialized nations of the world cannot in a sense be held directly responsible for shifting cultivation, the exploitation of trees for firewood and so on, other aspects of their conduct are directly relevant to the problem — such as their near monopoly of fossil fuels and their cornering of the means of fertilizer production.

Direct investment of western capital in commercial forestry operations in the tropics is, of course, a clear-cut influence. Japan and the EEC above all supplement home-grown building timber with ever-increasing amounts of hardwood from tropical sources. Though paper, rayon and other western consumer products are still mainly derived from the pulp of softwood trees grown in Scandinavia, North America and the USSR, there is a growing tendency the world over to turn even tropical hardwoods to these highly wasteful uses. Whatever the intended or ultimate use, *less than 10 percent* of the bulk of virgin tropical forest cleared in commercial operations actually winds up in a finished product.

Less than 10 percent. This is the harvest we reap from the total destruction of one of the few remaining unspoiled ecosystems on this planet. The remaining 90 percent (undergrowth, pulp residues, unmarketable types of tree and so on) is burned as waste. What is more, replanting and renewal are practices almost never used in the tropics. It is sadly true that dead forest is one of the few valuable resources many Third World countries can rely on to repay aid debts or buy industrial and agricultural hardware from Russia, Japan and the West. For the timber marketeer, tropical wood is a supremely easy deal, especially when worsening pest problems, soaring labor costs and machine overheads, and the eleventh-hour appearance of tough conservation statutes make forestry increasingly unprofitable in developed countries. The tropical forest is becoming an ever more attractive and inexpensive *objet trouvé* for the world timber and pulp industry to covet, but not to cherish.

Precisely what is the ultimatum which faces us, the world's most resourceful and persistent disease of trees? What in detail do we gain or stand to lose by deforestation?

It is relentless economic pressure which prevents us from striking a stable balance in our exploitation of trees, and this will be dealt with in a later chapter. Here I want to concentrate on the so-called "non-market" value of trees — the major impact that their decline, whatever its cause and extent, can be expected to have on the environment.

THE RESULTS OF DEFORESTATION

Air, water and fertile earth are the least replaceable contributions that trees make to life on this planet. This statement is no picturesque exaggeration. It is true that other plants are capable of making the same kinds of contribution — but in nothing like the same measure. The abundance and massiveness of trees add up to one mighty reason for their suitability. Again trees are intrinsically more adept in their use of raw materials than is the case with most other plants. That is precisely why they are so massive and abundant.

The interplay between trees and the air we breathe depends crucially upon a process unique to plants and essential to all life on earth — photosynthesis. Briefly, photosynthesis involves the conversion of the radiant energy of the sun into chemical energy. Using this energy, a plant can manufacture carbohydrates (sugars and starches) from different combinations of carbon, hydrogen and oxygen atoms. Land plants obtain carbon from atmospheric carbon dioxide (CO_2) gas, while hydrogen is derived from water (H_2O) molecules obtained mainly from the soil. The oxygen part of most water molecules is given off as a waste product.

The carbohydrates formed are then distributed around the plant. They are themselves stores of chemical energy. When reunited with free oxygen under certain circumstances, they can partly revert to their original components — CO_2 and water. In doing so, they generate energy that enables their remainder to link up with reactive mineral substances, especially nitrates and phosphates, from the soil.

Further energy transactions finally produce and activate proteins and other carbon-based substances that impel plant cells to grow and function.

The initial sunlight-transforming work is performed in plant leaves by chloroplasts — minute organelles harbored within cells in each leaf's middle layers. The well-known, light-sensitive green pigment, chlorophyll, is produced in the chloroplasts. Though chlorophyll is contained only in a minority of any one plant's cells, it is this which gives the earth's vegetation its overall green appearance.

Sunlight is in fact a mixture of several different kinds of energy, geared to several different intensities or wavelengths. We are normally conscious only of the fraction of sunlight that takes the form of heat and visible light. Other radiations have too short or too long a wavelength to register on our unaided senses, although radio waves, X-rays and so on are of course well known to us through our instruments. Visible light itself contains several segregated wavelengths of energy. Depending upon the nature of the surface they strike (i.e., its ability to either absorb or reflect them), these visible radiations are also evinced in our perception of the colors different objects appear to have. In a rainbow we can see the whole range of wavelengths in their fixed order from red to violet.

Chlorophyll absorbs many kinds of solar radiation, both visible and invisible, reflecting only the green light midway through the visible spectrum. Without the portion of this energy that chloroplasts subsequently contribute to photosynthesis, there would be no food and no breathable air on this planet.

Virtually all the free oxygen that exists in the atmosphere (about 21 percent of its total volume) originated through photosynthesis. Before primitive plant life evolved on earth, some thousand million years ago or more, the oxygen supply was disposed in chemical combinations with other elements, locked in composite forms useless to the development of free-oxygen-breathing life. Then ancestral forms of chloroplast evolved, probably first as self-sufficient free-moving organisms; they became parasites of other cells, and finally ended up as hereditary, acclimatized cell components — rather like latent viruses.

Only then did the atmosphere and the oceans come to acquire their full share of free oxygen. Some of it, in the uncommon form known as ozone, formed a layer at the atmosphere's outer rim and had the effect of shielding the earth's surface from drastic dosages of solar radiation. Plants no longer had to rely on water to provide a radiation shield and so were able gradually to colonize the land. Oxygen-breathing animals evolved and followed their source of food onto dry land.

It is true that plants also consume oxygen and produce carbon dioxide when their cells perform their various functions — including photosynthesis. When plants decay or take fire they again also use oxygen and give rise to carbon dioxide in considerable quantities. But, in life, plants usually absorb larger quantities of carbon dioxide and produce larger quantities of oxygen — through photosynthesis — than they ever tolerate during the reverse process of respiration. Only toward the end of their lifespan do plants respire and photosynthesize without producing a large overall oxygen surplus. And in any case most of the oxygen-consuming work done by plant cells is directed simply toward the perpetual increase and renewal of their light-sensitive, photosynthetic organs — their leaves — a kind of plowing back of capital, which gives still further dividends of precious oxygen in the long run.

The most characteristic features of trees, their height and rigidity, have evolved chiefly as ways to lift and spread leaf surfaces to intercept ever greater amounts of sunlight. Darwin likened trees to giant eyes, enormous stockpiles of light-sensitive cells. His diaries show that he sensed what several generations of biochemists have since confirmed — that trees are the world's main go-betweens in the transformation of universal radiated energy into life's necessities.

TREES AND THE ATMOSPHERE
Carbon Dioxide
At present the most frequently discussed aspect of the influence of trees on the environment is not their effect of maintaining the oxygen supply of our planet, but the part played by them in the cycling of carbon, particularly in the form of atmospheric carbon dioxide.

Carbon dioxide gas normally takes up around three hundred parts per million (p.p.m.) of the atmosphere's volume — about three-hundredths of one percent of the total. Accurate measurements and estimates of the global mass of airborne carbon dioxide are available only for the recent period from 1860 to the present. In that time, however, the level of carbon dioxide in the atmosphere has apparently risen from below 282 p.p.m. to over 330 p.p.m. This change is thought to have come about mainly as the result of burning fossil fuels (oil, coal, etc.). The output of these carbon dioxide sources is outstripping the absorbing capacity of carbon dioxide "sinks."

Forests are one of those sinks. They absorb carbon dioxide, weaving the gas's carbon base into a solid inert form — wood. But as we have repeatedly seen, the forests are diminishing — and so therefore is their efficiency and capacity as carbon dioxide sinks. Furthermore, because increasing quantities of forest are felled and allowed to decay, or are literally sent up in smoke, trees are being forced to act, as never before in history, as carbon dioxide producers. Tree respiration and the normal decay of dead forest vegetation was, of course, always a source of the gas — but a slow-acting and redeemable source. Slash-and-burn cultivation and the use of trees for fuel, however, now rival fossil fuel burning as major factors in carbon dioxide overproduction. If the surplus continues to increase at its present rate, carbon dioxide in the atmosphere promises to reach levels of over 650 p.p.m. in the next fifty years(14).

What effect, we have to ask, could levels like these be expected to have on the living world? The most disconcerting among many disturbing answers lies in carbon dioxide's so-called "greenhouse effect" on the air temperature near the earth's surface.

When solar radiation reaches the surface of our planet via the atmosphere, some of it is absorbed as light by photosynthesizing plants and as heat by all kinds of land and sea surfaces. Most of it, however, simply bounces off the planet in the form of heat energy and heads straight back into space. The presence of carbon dioxide in the atmosphere, however, delays this heat loss. The physical structure of that gas is such that it allows light waves to pass through it at a greater rate than it lets through heat waves. In other words, because of carbon dioxide, some of the sun's energy, which arrives as light and is turned into heat at the earth's surface, radiates off more slowly than it comes in.

The earth's temperature and climate, therefore, are governed to a great extent by the relative surplus of carbon dioxide in the air. Average temperatures worldwide have recently been increasing, as it happens, in step with the increase in the atmosphere's carbon dioxide content. Other factors, however, can also affect this situation. The buildup of atmospheric pollution may, for example, be exerting a temperature-*reducing* effect in some parts of the world.

Such complications aside, the overall result, say, of a doubling of the greenhouse effect would be a dramatic rise in the world's average temperature by some five to six degrees centigrade. This rise may seem small. But it would in fact suffice to cause parts of both polar ice caps to melt, raising sea levels everywhere by several dozen meters. This rise would be sufficient to engulf major centers of human population situated on major rivers and coastlines. At the same time, the climate of equatorial areas would become hostile to most forms of life.

A further, even more significant, consequence of the production of extra oceans might paradoxically be an eventual drastic cooling of the earth's surface. For unlike ice, liquid water can absorb carbon dioxide very readily. So it is quite feasible that a great increase in the ocean surface could abolish the greenhouse effect almost completely, by taking carbon dioxide out of atmospheric circulation, or by greatly increasing the earth's overall tendency to reflect heat back into space.

As long ago as 1861, the physicist John Tyndall advanced the theory that fluctuations in the carbon dioxide component of the atmosphere might be a major cause of ice age climates. Most modern theories of climate history do not contradict Tyndall's physics — they merely express differing notions as to how atmospheric imbalance is brought about in the first place. Fossil fuel burning and deliberate deforestation, arguably the main reasons behind the present-day carbon dioxide surplus, have of course no obvious counterpart in prehistory. Commentators therefore look farther afield and aloft to find the factors that altered the atmosphere's capacity to control its own heat in the distant past, and so perhaps gave rise to ice ages. Cyclical spates of volcanic activity, periodic changes in the earth's orbit and a range of solar and cosmic phenomena have all been proposed as figures in the equation. While I am not attempting to dispute those theories absolutely, I suggest that pathological deforestation (an age-long losing battle between slowly evolving trees and quickly evolving tree parasites, resulting in periodic disease epidemics which may, at crucial moments, have upset the global carbon dioxide budget) could be a significant contributory factor. This would be prehistory's equivalent to the more self-evident causes of today's carbon dioxide boom. The idea is speculative, but no more so than, say, the notion that ice age conditions occur when our solar system periodically takes a dive into swathes of cosmic dust which produce, so to speak, a galactic greenhouse effect(17).

Whatever view they favor, scientists now generally agree that the present buildup of carbon dioxide in the atmosphere could, if it were to continue without restriction, spell catastrophe for our planet within less than a century. Exactly what to do about it must remain in question until we have more exact knowledge of the identity and relative importance of carbon dioxide sources and sinks. But time is not on our side.

Oceans are carbon dioxide's most important known sink. About thirty thousand million tons of carbon in the form of carbon dioxide are reckoned to dissolve into their waters each year. Nearly the same amount is supposed to leave them yearly through evaporation. But an estimated reserve of some forty thousand billion tons constantly remains in solution and in the sediments on the ocean bed.

The amounts of carbon dioxide thought to circulate through the earth's biomass seem modest in view of the ocean's appetite. But they are, of course, just as essential to the overall equation. Plants are believed to absorb about twenty thousand million tons of carbon dioxide a year through photosynthesis and to repay comparable yearly amounts through respiration and decay.[1] The carbon held in reserve in forests in the form of wood and woody topsoil has been estimated at 1.6 billion tons. Other, more ephemeral, types of vegetation (including green plankton in the oceans) presumably conserve smaller, but still considerable amounts.

I have already mentioned that most of the atmosphere's current carbon dioxide surplus is, by conventional accounts, a byproduct of the burning of fossil fuels. Estimates of the amount of the gas that issues from this source range from two thousand to five thousand million tons a year. The lower suggested rate is still, by conservative calculation, about a hundred thousand times faster than the rate at which carbon in the form of new fossil fuel materials is deposited in the ground. Such a vast discrepancy cannot by any stretch or neglect of the imagination be viewed as part of a balanced cycle.

But the part of the carbon dioxide surplus that originates from fossil fuel burning is even then, it seems, not the whole problem. Deforestation also plays a part by placing a heavy strain on the otherwise normal tendency for forest and ocean to show a neat balance between intake and output of atmospheric carbon. The large amounts of wood carbon that we incorporate into buildings, thus postponing its decay (i.e., its potential to form atmospheric carbon dioxide), are insignificant in relation to the amount of carbon dioxide we generate simply by burning our trees or by allowing the waste products from our exploitation to decay.

An international conference of forestry concerns held recently in Berlin discussed a figure of 9,200 million metric tons as the likely annual total of fossil fuel plus deforestation carbon now entering the atmosphere as surplus carbon dioxide(14). The world's carbon dioxide sinks are apparently working overtime to absorb some of the surplus, for only about 2,500 million tons a year currently remains in the atmosphere. If these measurements are accurate, where does the rest go? If the dwindling biomass, represented mainly by forest, is now more of a source than a sink, then the oceans (contrary to available data) must surely be doing most of the extra absorbing — unless other unrecognized sinks exist. The most certain of all the relevant observations is this: that the ocean, whatever its present capacity to absorb the gas, must become less and less able to act as a sink as time goes by. Carbon dioxide is the gaseous form of carbonic acid; the more the ocean waters absorb, the more acidic they become and — it so follows — the less absorbent. Indeed, their capacity in this respect can reasonably be expected to halve every twenty-five years under present conditions.

How can a dangerous bottleneck of atmospheric carbon dioxide be avoided? The obvious answer is, by curbing the use of fossil fuels, but it is plainly an answer that today runs straight against the grain of real politics and economic habit. The likelihood that world reserves of oil and gas will run out in forty or fifty years' time subtracts very little from the problem, for oil and gas are far from being the most potent sources of carbon dioxide. Coal is a much more productive source and present reserves of coal will last, according to most experts, well into the twenty-third century, even assuming a 4 percent annual increase in world coal consumption. Such an increase would, however, have produced enough of a carbon dioxide surplus to make the climate hostile to agriculture in most parts of the world long before the end of the twenty-first century. Only the development of alternative energy sources can forestall this threat, unless improved methods — artificial or natural — of sinking carbon dioxide can be found.

A practical defense against a ruinous carbon dioxide surplus would of course be to secure the future of forests to act once more as a sink, rather than as an overall source of the gas. This naturally means curtailing deforestation and counteracting it by afforestation on a suitable scale.

The bulkiest current obstacle to the curbing of deforestation is the pressing need to increase the world's complement of agricultural land — to feed the world's ever-growing population. It is far easier to destroy forests than to rehabilitate deserts for this purpose. Unless the human population explosion can be abruptly limited, there is no apparent end to this paradox. It is only realistic to suppose that deforestation will continue until birth control is universally practiced and the world's material resources (particularly the supply of soil fertilizers) are evenly distributed among its nations. If it were true to say that agricultural land could substitute for forest as a regulator of the environment, there would, of course, be no problem. But this possibility is unlikely.

In today's world we are faced with the need to find a compromise that will allow forests to decline yet still let them allow us to survive and develop. The least unsatisfactory form such a compromise might take could be that of "controlled" deforestation. Managed strategically, a few trees can theoretically do the environmental work of many, as far as gas exchange with the atmosphere is concerned. Simplistically speaking, free-standing or woodland-edge trees can interact with the sun and air more efficiently than trees at the heart of a dense forest. In some situations, it may therefore be possible to thin and redistribute large woodlands into smaller units and actually *increase* their carbon dioxide intake.

Tropical rain forests could not be parceled up in this fashion and retain their identity. Their ecology is too complex, their components too interdependent. Tropical forest ecosystems are liable to disintegrate completely as a result of quite minor reductions in their absolute variety or size. It is impossible, on the basis of present knowledge, to measure and manipulate these critical limits.

In temperate lands, however, there is a good deal more room to maneuver. Most temperate forest trees can and often do stand in small groupings without losing their character as habitats for a variety of other organisms, and without failing in their effectiveness as key environmental governors and aids to agriculture, even without becoming disqualified as commercial concerns or recreational areas. This is, again, necessarily a slightly oversimplified view, for there are environmental imperatives (e.g., the prevention of flood erosion in hilly regions) that still require the presence of massed forests in certain areas. There is, nonetheless, a general case to be made for the splitting up of some large forests into a multitude of smaller units, along with the redesigning of existing temperate farmland to incorporate more trees. Such a redistribution of tree populations would entail a willingness on the part of entire societies to accept responsibility for the upkeep and renewal of woodland and the control of tree disease.

Though the sinking of carbon dioxide is likely to be a major object of forestry exercises in the future, it certainly comes rather low in planners' present lists of priorities. In the files kept by the US government's energy agencies there are contingency plans to correct atmospheric carbon dioxide surplus by legislation for the use of wood rather than steel or concrete in the construction industry (thus stimulating afforestation and locking up wood carbon out of harm's way) or even, as a last resort, by continually growing noneconomic forests to the peak of their ability to photosynthesize, then axing them and dumping the trunks down worked-out coal mines! These plans will presumably be left on the shelf until the world shows unmistakable signs of suffocation.

Meanwhile, the progress of bulk afforestation in developed countries continues to depend almost exclusively on market criteria only. Even though forests can be planted in bulk quite cheaply on poor land unfit for farming, large-scale forestry in most developed countries is not much of a commercial proposition except in regions that still have a large natural reserve of dense woodland. For timber is a crop that takes from four to fourteen decades to become harvestable, and is extremely prone to lose its profitability through inflation, even leaving aside the perpetual risks of fire, disease and so on. In many cases, large-scale commercial forestry survives only at the behest of state subsidies. Those subsidies are officially justified by reference to rather debatable concepts such as the amenity value of bulk woodland and the tourist attraction it exercises. Yet these concepts apply just as much, if not more, to small-scale, labor-intensive forestry. Bulk commercial afforestation, geared to an enormous capital investment in expensive machinery, is, moreover, likely to become something of a white elephant as the raw materials for the production and maintenance of giant hardware become scarcer and costlier.

Oxygen

Though certainly controversial, the balancing of the world's carbon dioxide budget is by no means the only important effect trees have on the environment. The circulation of oxygen in the atmosphere is another phenomenon in which trees, through photosynthesis, play a key part. For every unit of carbon dioxide they absorb, trees can release anything up to four units of oxygen into the air.

Like carbon dioxide, oxygen is difficult to monitor globally. It evens out at around a fifth of the atmosphere's volume, but measurements often differ slightly from site to site. A local drop in the oxygen level is an anomaly quite frequently observed. Though there is no theoretical ground for supposing that oxygen might become unpredictable in its global distribution, that is what some of the practical evidence seems to suggest.

Writing in 1972, Curry-Lindahl(3) said he knew of only five scientists, out of the many hundreds who had commented in print on the future of the world's oxygen supply, who were prepared to state categorically that the supply has no foreseeable limit.

The necessity for drawing global conclusions from a few local samples lends a rather farcical inconsistency to the literature relevant to the atmosphere's composition. Basic premises tend to be replaced overnight by their opposites and the frill of theorizing that accumulates around one set of figures often has to be laboriously unpicked or tacked on to another. For example, it was once thought by most authorities that green plankton in the oceans was the main source by far of recycled free oxygen — anything up to 70 percent of the total supply was attributed to aquatic vegetation. Subsequent investigations have tended to show that plankton is responsible only for about 30 percent of the free oxygen supply — the rest circulates through land vegetation, presumably mainly through trees. If this last view stands firm, then deforestation must certainly have a drastic effect on the availability of atmospheric oxygen as well as that of carbon dioxide. Figures given for carbon dioxide surplus resulting from deforestation imply an atmospheric oxygen loss of even greater proportions.

Could a shortage of atmospheric oxygen actually already be a reality? In the absence of adequate geophysical data, we can perhaps look to the medical record for possible clues. The increasing incidence of heart and lung diseases among humans in many different parts of the world, for example, is a problem conventionally attributed to industrial pollution and the extra stress that modern living is supposed to exert on overfed and underexercised towndwellers. But there are signs that the problem is not limited to societies where those factors are present. Other, more universal stresses seem to be in operation. Is instability in the level of atmospheric oxygen one of them?

No statistic suggestive of such a link has ever been published. But if one remembers that the link between tobacco smoking and disease took decades of research to establish, it makes sense to assume that an unhealthy shortage of atmospheric oxygen need not become recognizable to medical science until it had already run to serious extremes. If the threat of oxygen impoverishment in the atmosphere is debatable, surely then the least we can do is to debate it to our entire satisfaction — and certainly before we continue to sanction the destruction of the world's most efficient oxygen producers, the forests.

Water

Water is yet another vital commodity which trees, more than any other kind of organism, trade with the environment. Their trade is, so to speak, both wholesale and retail. For example, highland trees act as a wholesale outlet for cloudborne water (mostly the product of ocean evaporation) by "roughening" the air above and around them. The extra atmospheric turbulence they generate wrings extra rainfall from passing cloud masses. In western North America, around 90 percent of the water that is used for consumption in households, industry and agriculture originates on forested watersheds.

The expression "watershed" describes the stable level of water saturation that exists beneath any suitable terrain, kept within limits by a "table" of watertight bedrock. When full, mountain and upland watersheds spill over into the rivers and streams which irrigate lowland areas. Forests do a great deal to keep this supply constant, not only by encouraging rainfall at source but also (in their "retail" role) by regulating the rate at which water descends to the sea. Forested slopes and riverbanks are protected from erosion by an umbrella of foliage and a maze of tenacious, interlocking tree roots. The forests thus guard against the ravages of floods and landslides. More than this, the removal of forests could "unplug" successive watersheds and impair their ability to supply the land with constant amounts of water at dry times of the year. In areas of high snowfall the shade of forests also delays the spring thaw and ensures that the melting snow is transformed gradually into groundwater instead of spending itself in wasteful and destructive floods.

And trees, wherever they grow, show stubborn resistance to drought, reaching deep into the ground with their roots, and conserving sizable stocks of moisture in their vascular systems and in the ground they shade. Thus they provide a haven of moist air and soil for animals, and for less hardy plants. This "oasis effect" can be plainly seen at work in agricultural land laid bare by drought. Crops and weeds growing in the close vicinity of trees are often the only greenery the fields can show and cattle and flies press into the shade of any available tree or woody shrub, no matter how meager.

More generally, trees also reduce evaporation losses from the soil by acting as windbreaks. Marquardt's(15) studies in Germany have shown that the planting of trees and hedges around small grain fields

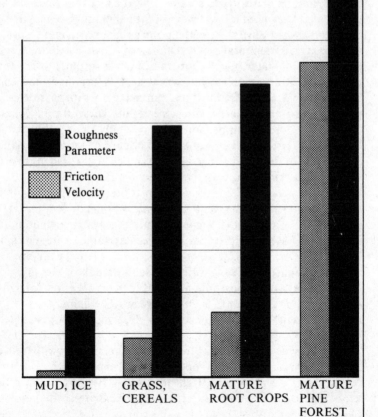

Figure 2
*Relative effect of different surface features on redu-
cing wind speed near the ground and setting up patterns of air turbulence likely to enhance rainfall above
an otherwise flat terrain. Measured in terms of friction velocity which is in turn dependent upon roughness parameter. For a further explanation of these
terms see M.A. McIntosh,* Meteorological Glossary,
London, HMSO, 1972.

Roughness
Parameter

Friction
Velocity

MUD, ICE   GRASS,   MATURE   MATURE
            CEREALS   ROOT CROPS PINE
                                 FOREST

can increase crop yields by 20 percent, simply by helping to preserve the ground from the drying effect of wind. Similarly high yields would, he maintains, be obtainable in unhedged fields only if the normal rainfall were to be supplemented there by an extra one-third.

Trees actually consume, for their individual needs, far more water than do other kinds of plants. Most of their intake, however, is not put to work in growth processes or photosynthesis but is simply shuttled through the tree and transpired into the air through the leaves. An area of ninety square meters of pasture or wheatfield transpires, for instance, about fifty tons of surplus water per year, while a single deciduous tree can transpire as much as three-quarters of a ton *per day* during summer weather. Trees growing in agricultural situations are therefore often viewed by shortsighted farmers as weeds, stealing much-needed water from their cash crops.

The paradox arises quite simply from the fact that trees are big organisms. They need a lot of water if they are to survive but, on the other hand, they generally use their lion's share much more frugally than do most agricultural crops. Alfalfa, for instance, requires about 800 pounds of water (mostly for transpiration surplus) to produce one pound of dry matter, while wheat needs 450 pounds and corn 350 pounds. A pine tree, however, can make do with as little as 50 pounds of water to achieve the same result. These are, of course, extremes and such comparisons do not apply universally. Apple trees require in summertime about 500 pounds of water to produce each pound of their dry weight, although much of it is used in the production of a watery fruit which has food value.

Exceptions such as the apple do not mar the general rule that trees as a whole consume water more economically than do fodder, cereal and root crops as a whole. Figures showing the water content of different plants' leaves, expressed as a percentage of their fresh weight, confirm this truth. The leaves of a lettuce or dandelion plant are, for example, about 85 percent water, those of an oak or elm growing under similar conditions only about 65 percent. Figures for grass and root-crop leaves average 70-80 percent. The living parts of the woody stems of trees have, not surprisingly, a very low water content — typically 50 percent or less.

It might be said that most of the field plants compared with trees in these examples are valuable food crops and should, if it came to a choice, have a certain extra right to scarce water resources, in the sense that they are more directly useful than trees to humans. Against this narrow point of view must be weighed, first and foremost, the special considerations trees merit as general influences on the environment and atmosphere. But in any case there are plenty of much more specific considerations that amply justify the existence of trees in terms of harvestable gain. Some of them will be dealt with later in this book. But they should not divert attention from the supremely important concept of the overall effect trees have on the living world's most basic necessities.

Soil

Trees are closely linked both to the formation and the conservation of soils.

The fertility of virgin soil depends greatly on its humus content. Humus is dead organic matter, mainly of vegetable origin. Fungi, insects, worms and microbes feed on this, gradually rendering it into byproduct chemicals essential to the survival of still-living plants. The chief and most stable component of humus is lignin and most of the lignin in the soil has been derived, at one time or another, from the decomposition of trees or undergrowth associated with trees.

Besides supplying raw material, trees also naturally control the progress of soil formation by governing the amount of light, heat and moisture reaching the ground in their neighborhood. Thus they program and enhance the output from a medley of decomposition activities. Trees also help to provide the inorganic basis of soil. Their roots, rummaging deep into the ground for water, probe the rocky subsoil and divide it into smaller fragments — a step toward its becoming grains of topsoil. Whatever happens, soil formation is a slow process; a topsoil one foot thick may take five hundred to a thousand years to develop and to achieve optimum fertility. Human cultivation, however, can undo this work in a few growing seasons by limiting the extent and diversity of the decomposition process and by laying the soil open to erosion.

Industrially produced soil fertilizers can make good the loss of fertility in cultivated soils, but their supply is in any case limited. Already fertilizers are, weight for weight, one of the most expensive commodities that world trade handles. Furthermore, their persistence in the soil is short lived — they are all too liable to drain away from the land before their full effect is felt and to percolate into rivers, where they cause a variety of pollution problems. The humus they in some ways replace is, on the other hand, capable of inhabiting and fertilizing the soil over a much longer term, meanwhile lending it a firm texture and helping it retain moisture.

The current imperative to produce food plainly cannot always wait for nature to take its course where the replenishment of the soil is concerned. But it is hard to see how, in future generations, the world's soil is going to remain fertile and plentiful in the absence of a ubiquitous woodland element in the vegetation. No matter how sophisticated crop husbandry may become, the raw materials of soil formation will always be subject to a law of diminishing returns unless they are provided for in the present. In this sense, the woodland of today is the agricultural land of tomorrow — just as we now owe the viability of present-day soils largely to forests which overshadowed them long ago.

The conservation of existing supplies of soil is firmly tied to the conservation of trees. Of that much we may be sure. Soil is, after all, a resource (and a limited resource) in its own right. Every year, thanks to human activity, billions of tons of topsoil are unnecessarily washed into the oceans and lost to human posterity. The evolution of fresh topsoil is far too slow, under present conditions, to replace this loss. Yet the beleaguered offspring of the forest which sponsored the formation of the majority of present soil stocks, continue steadfastly wherever possible to protect soil, by counteracting erosion and stemming the day-to-day runoff. A study undertaken in the USA in 1969 demonstrated that the sediment runoff during rainstorms at highway construction sites stripped of all vegetation was, on average, ten times greater than from grassland and *two thousand times* the loss from forested areas. In deforested areas, then, the ability of the land to retain its soil cover can be reduced by 90 percent.

RESERVATIONS AND CONCLUSIONS

It would be wrong of me to present the comparisons made in this chapter without drawing critical attention to the possible and actual shortcomings of the experimental studies on which they are based. To any statement which depicts trees as star performers of this or that environmental function, the proper response is "which trees?" and "growing where?"

Unlike "corn" or even "grass," the word "forest" does not describe a class of closely related organisms with a roughly similar biology. The evolutionary lineage of trees is thoroughly varied. The plant communities we loosely call forests are infinitely different, complex collections of living things, plant and animal, whose combined impact on the environment can only usually be measured in vague and gross generalizations. Even the finest measurements will yield results valid only for one patch of forest at one particular moment.

We have to understand that a forest is a process as much as a thing and that its status in the environment changes, gradually but inevitably, from minute to minute. The same is mildly true also, say, of a cornfield. But it is nevertheless feasible to measure the functioning of a square foot of cornfield and, with the help of mathematical and climatic tables, arrive at conclusions which will hold reasonably true for the physical properties of cornfields the world over. But when attempts are made to fit forests into a credible overall statistic, the material is nothing like so obliging. Different attempts produce widely different results at different times and in different places — and not through errors in observation. The differences are differences of fact. It is hard even to generalize about the biology of two different kinds of trees, let alone to draw comparisons between trees and other plants as a whole.

Just how difficult the theoretical and conceptual problems are can be shown by looking at the exchange of gases between trees, other plants and the atmosphere. We need to compare the rates of photosynthesis (intake of carbon dioxide, release of oxygen) with that of respiration (intake of oxygen, release of carbon dioxide). Figure 3 overleaf examines the rates of these processes for six different trees and crop-plant communities.

In the case of the tropical jungle, conditions for photosynthesis are optimal all year round. Other plant communities which inhabit regions of seasonal climate do most of their photosynthesizing in a frantic summertime spurt, then moderate their efforts or disappear from the reckoning altogether at other seasons. So, if the *annual* net turnover of photosynthesis and respiration products were depicted for the different communities featured, they would have to be ranked almost in reverse order to that shown in Figure 3.

Again, the age of the community under study is crucial to the balance it shows between carbon intake and output. The forty-year-old jungle in our example is close to maturity in the sense that its respiration peak is quite close to its photosynthesis peak. Other, younger rain forests would give a quite different statistic. Further uncertainties arise from variations in wind movements through the foliage of a mass of plants and from differing amounts of light and water at the service of photosynthesis in different situations.

Once more, the tropical forest is special on this score, for it is characteristically segregated into two, three or more different stories or layers of vegetation, crowned by a dense upper canopy. Trees belong-

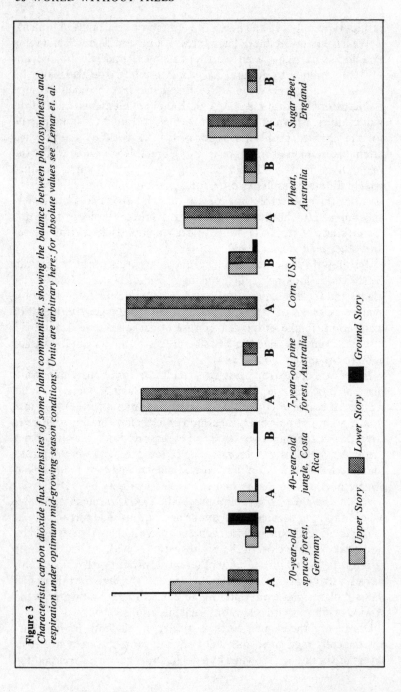

Figure 3
Characteristic carbon dioxide flux intensities of some plant communities, showing the balance between photosynthesis and respiration under optimum mid-growing season conditions. Units are arbitrary here: for absolute values see Lemar et. al.

ing to successively lower stories are habituated to successively smaller supplies of light and air. In the bottom stories, they may obtain most of their carbon dioxide requirement from the updraft of respiration products issuing from decomposition activities on the forest floor. Some forest floor plants can photosynthesize in light no more intense than that of a half-moon! Undisturbed tropical forests might therefore be said to have rather a "sealed system" of gas exchange. The uppermost foliage draws most of its carbon dioxide needs down from the atmosphere. The canopy is usually so dense that the photosynthetic surface it presents to the sun is virtually one-dimensional, like a meadow on stilts.

Though other kinds of forest can show a marked partitioning into stories, they are generally more prodigal in their interplay with the outside atmosphere, usually because their leaf area is greater per area of land, or more three-dimensional in scope.

Turning to the crop plants instanced in Figure 3, their photosynthetic efficiency looks quite respectable in relation to that of the forest samples. True, they are less efficient in other ways and their effect on the atmosphere is shorter lived, but their repeated contribution is guaranteed by their importance in agriculture. Furthermore, certain tropical grasses are known to maintain very high rates of photosynthesis throughout the year, outdoing even rain forests in this respect at least.

When such reservations are voiced, we are again left with the awkward conclusion that comparisons between this forest or that forest, or between this forest and that cultivated plant cannot consistently be made in terms of the relative value of their contribution to a single environmental ideal. In other words, we cannot be absolutely certain that, if we replace natural forests by agricultural fields or commercial plantations, we irrevocably upset the balance of the environment. The knowledge now available simply falls short of final scientific wisdom, wisdom that might be reflected in the policies governments pursue toward forests. To enumerate the uncertainties involved is not, however, to agonize about them, but to attempt to share them more widely as potent parts of the argument *against* deforestation. In his *Provincial Letters*, the philosopher Blaise Pascal declared: "We must be able to doubt when necessary, to be certain when necessary, submitting ourselves when necessary. He who does not act thus understands not the force of reason."

We know enough in a general sense to be aware of the nature of the threats deforestation poses to the well-being of the environment we and all creatures share. There are limits to the time we can take to resolve the doubts surrounding our fullest understanding of the dangers which present knowledge foreshadows. Surely it makes sense now to submit to the need to prevent deforestation with all the means and arguments at our disposal, until we can be certain about what we are doing to our planet when we remove from a single tree its right to a space here.

Or must we, as so often in the past, find out the hard way?

3 Portrait of a Tree

"The moan of doves in immemorial elms. . . ."
ALFRED LORD TENNYSON
"The Princess"

"The tree which moves some to tears of joy is in the eyes of others only a green thing which stands in the way. Some see nature all ridicule and deformity and by these I shall not regulate my proportions, and some scarce see nature at all. But to the eyes of the man of imagination, nature is imagination itself. As a man is, so he sees. As the eye is formed, such are its powers. You certainly mistake, when you say that the visions of fancy are not to be found in this world."
WILLIAM BLAKE
Letter to Dr. Trusler

RATIONALE

TREES HAVE HISTORIES in precisely the same sense as we ourselves have a history. In their case, as in ours, ages of supreme prosperity alternate with periods when extraordinary difficulties are endured — invasions by competitors, sudden disasters, and hostile alterations in general conditions. Such changes of fortune, as with ourselves, leave their mark on individuals and whole populations, conditioning the unique ways in which they respond to their world and objectively defining that world in time and space.

Though the history of most tree species is several million times longer than human history, our domination of most of the earth's dry land has brought us lately more and more into competition with trees — while nevertheless not actually altering our status as one of their many natural dependents. The result is that their recent history has become as eventful and rapidly changing as our own. Human wars, for example, and civil power struggles over the last two or three hundred years have worked profound effects on tree populations. Even at humanity's most harmonious moments the relationships between trees and men is not better than a stockmarket affair, and at worst is an outright, active jockeying for territory.

It is now reasonably true to say that certain kinds of trees flourish on the planet because commercial markets require them to do so, while other kinds grow scarce because their human usefulness is currently considered not substantial enough to command our cooperation in their struggle against natural and manmade hardship. Or sometimes simply because we value their territory or material more than their useful living presence.

But our economic valuation of trees — trees in general as well as particular kinds of tree — is no more consistent than our economic valuation of anything else. The net market value of any commodity changes dramatically according to demand, and always will. Unlike other commodities, however, a supply of tree products for specific purposes cannot be manufactured, diversified or remaindered overnight in response to rapid changes in demand. On the whole, trees remain mainstays of human commerce and industry simply because, despite our misuse, they are still there to be exploited. They are still (though not for much longer) plentifully available in nature, so their exploitation (as opposed to their conservation) requires little investment of capital or research and development funding. This goes for the tropical subsistence farmer just as for the western timber trader, but the latter's approach is easier for us to analyze.

For general heavy purposes, such as provision of construction timber or the manufacture of paper, tree species are fairly interchangeable — at least in the sense that exhausted home-grown supplies of useful kinds of tree can readily be supplemented (in peacetime) by imports of similar trees growing abroad. To this extent, world demand for marketable trees responds to supply, rather than the other, more usual way around. Where specific valuable commodities — quinine, for example — can only be derived from particular kinds of tree growing only in particular areas, then extra time and money are spent on protection and enlargement of the source of supply. In more general markets, the bulk and continuity of supplies can often be achieved with a minimum of cultivation. It is therefore extremely rare for any nation to develop a successful home-grown forestry or heavy forest products trade from scratch. Large forestry enterprises are nearly always founded, as suggested, on the status quo of an enormous and seemingly inexhaustible stock of natural forest, plus a hungry export market.

How do these generalizations square with those quite different ones made in the foregoing chapter? There we were concerned to understand the massive support trees lend to the environment which we share with them, and some of the enormous risks that deforestation poses to the health of the world at large. In this chapter, I want to approach the clash of interests (health of the planet versus economic exploitation) by scaling the argument right down to the intimate consequences that the population decline of one kind of tree — the elm — is having on the quality of human life.

The familiar elm is a textbook case of a tree which has, in the past, served an intriguing variety of human uses, uses no other available tree could have served better. There is no reason to doubt that it could also serve essential uses in the future, if it happened to be around in quantity at the time. As it happens, some of the tropical timbers nowadays used in manufacturing processes which formerly depended on elmwood as a staple are just as likely to disappear from the world market before the end of the century. This disappearance will occur either through overexploitation or — paradoxically — thanks to strict protection. For these and other reasons, it is realistic to suppose that a completely unprecedented need could someday arise that would suddenly place a high market premium on elms. Who, for instance, could have guessed in advance that elm timber would become an important raw material of the early automobile industry in the USA? Certainly not the men who planted the trees.

The fact remains that the present market value of various kinds of elm timber, though higher than many people realize, is not high enough anywhere to win a direct anti-DED campaign a place in development plans at national level. Planners have trouble enough forecasting day-to-day shifts in the value of ready-to-hand commodities. To invest a large fraction of national funds in *ad hoc* elm-protection or replacement programs, solely on the off chance that elms will have an extra-high cash value on an unforeseeable market fifty or a hundred years hence, is plainly out of the question in strict business terms. Even if one knew for certain that some such market would eventually evolve, fifty to a hundred years is in any case the time it would take to rear a viable stock-in-trade.

Absolute, current market value cannot be ignored as a decisive factor in the securing of tree species from neglect or active destruction. It is the one consideration that can today be guaranteed to appeal to the sympathies of policymakers in western countries and their economic satellites. But the application of market values to conservation problems (typified by the plight of the elm) appears especially inappropriate, even flippant.

A great deal of time and ingenuity has lately been spent by conservationists in attempting to place a cash value on the nonmarket benefits of unharvested or unharvestable trees. The purpose of the exercise is clear. It is to save influential planners the trouble of making an individual or an alive response to conservation arguments, by couching those arguments purely in balance-sheet terms. Yet most of the nonmarket benefits we derive from a tree like the elm are far too subtle or perennial to seem important on a balance sheet. Any cash estimate of their worth can easily be matched by a cash estimate of the nonmarket value of the pastures, sidewalks or playgrounds that might fill the spaces elms now occupy.

Yet surely nonmarket considerations must be evoked in their own name, wedded to their own rightful logic, when a case is to be made for rescuing European and American elms from the limbo they are now heading toward. Those considerations range from the immeasurable value of elms as parts of landscapes, to accountable facts centering on the proposition that it is often more costly to dispense with dead than to protect living elms. In between, a wide range of ecological, agricultural and economic arguments can be identified, which show that the decline of the elm is a far more serious matter than most published accounts suggest. Some of these arguments will be woven into this chapter.

The record of past and present links between humans and elms is fascinating, but not exceptionally so. For a study of the history of our relationships with many other kinds of tree would be just as eventful and rich in interest. But our connection with the elm is, of course, at a stage which is especially diagnostic of the way we currently appreciate (or fail to appreciate) trees of all kinds. It is improbable that we can tackle the complex and enormous task of protecting the world's trees until we can learn to love and cherish the trees in our own backyards.

National initiatives to conserve elms are limited mainly to research programs for breeding disease-resistant replacements for today's DED victims. This is a shrewd yet fallible approach. For, so far, fifty years of breeding programs have failed to produce a "superelm" and there is no certainty whatever that fifty more years will produce one. The only large-scale national attempt to conserve elms in the field — the DDT-spraying program initiated by the US Department of Agriculture in the late fifties — was discontinued, rightly but ironically enough, on conservation grounds. In her classic work *Silent Spring*(2), Rachel Carson made a powerful case against the environmental hazards of DDT. Popular outrage did the rest.

Though insecticide sprays have long since been superseded by environmentally safe anti-DED techniques, control fieldwork is now almost everywhere left to the efforts of local amenity organizations, commercial pest-control operatives or individual landowners. Despite the often conspicuous success of these efforts on a parochial scale, the resources needed to translate the approach into national elm-protection programs are no longer forthcoming. The policy priorities set by conventional wisdom therefore must be reorganized to some extent before saving elms can become a national concern in each country. The same muscle of public opinion that wiped out DDT-spraying of elm trees is now sorely needed to enforce new and sensible techniques of DED control on an effectively large scale. If this end could be achieved, it would raise conservation hopes everywhere, far outside the boundaries of countries where elms happen to grow. For the economic habits and arguments that stand against any intention to save elms are miniature replicas of the obstacles that confront grand attempts to curb deforestation and other devastation on a planetary scale. Anyone seeking some indication of the way governments will deal with deforestation during the next fifty years need look no further than at the way we cope with DED during the next five.

Now follows a portrait of a tree — a basis for reflection on things we stand to lose, if the notion that research alone will save the elm is not borne out by events.

THE EVOLUTION OF THE ELM

There are between thirty and forty-five species of elm extant — the identity of several species being currently under revision. Cultivation by man has also given rise to many hybrids of untraceable origin. All elms belong to the genus *Ulmus*, a part of the botanical family Ulmaceae. That family also includes uncommon trees like the zelkova and the better-known shrub *Celtis* or hackberry. *Ulmus* species are extremely varied in appearance, ranging from tall and heavy trees like the American, English and Dutch elm (*Ulmus americana, U.procera* and *U.hollandica* respectively) to more delicate, almost birchlike trees like the Siberian elm (*U. pumila*) which have much smaller, narrower and more serrated leaves.

Elms can be distinguished from most other trees by their asymmetric leaf base. Thus:

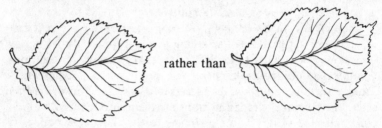

rather than

There are other anatomical features peculiar to elms, but this one is the most easily identified. Elms are distinguished from other members of the Ulmaceae mainly by their fruit, a flying-saucer-shaped samara, which is analogous in structure to stone fruits like cherries, with this difference: the fleshy sphere that surrounds a cherry drupe appears in elms as a flat, dry envelope adapted as a wing to aid the seed's distribution by wind.

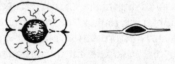

Elms occur throughout the temperate and in many of the subtropical parts of the northern hemisphere (see Figure 4).

Their major centers of population in modern times are northwestern Europe, eastern North America and northern Asia. Their former distribution was, as the map shows, more extensive still, taking in several locations high in the Arctic Circle.

Elms first evolved into the form we recognize today during the Cretaceous Period, about a hundred million years ago. At that time the continents had roughly the same boundaries they have now. But they were then still relatively close neighbors on the globe, separated only by narrow straits of shallow water. Also, the areas where elms are now chiefly concentrated on the globe were situated not high in the temperate zone but in the subtropics. The common ancestral stock of modern elm species was split up and carried to cooler quarters by the increasing separation of the continents. The great variety of new and strange environmental pressures occasioned above all by this reshuffle began to give rise to species that necessarily diverged in shape, size and biology from the ancestral type.

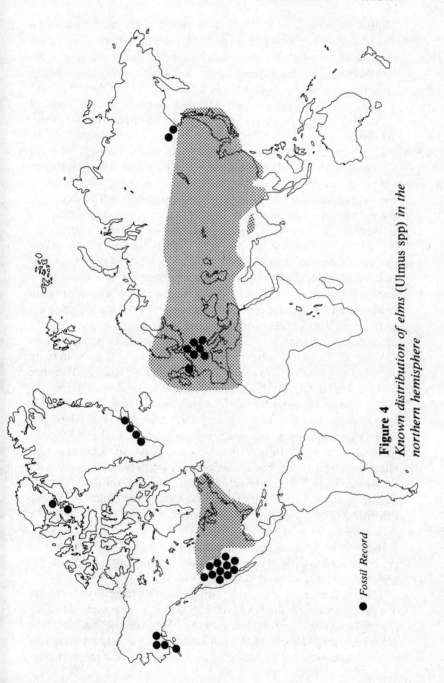

Figure 4
Known distribution of elms (Ulmus spp) in the northern hemisphere

● *Fossil Record*

Elms evidently adapted well to the new environments foisted on them by these age-long changes. Their northernmost representatives, in what is now Greenland for example, failed to persist. But elsewhere the divorce of elms from their ancestral climates (where competition with many other trees had been unrelieved by geographical separation) worked to their overall advantage. The influence of natural selection now gave rise to kinds of elm specialized yet versatile enough to tolerate a wide range of more-or-less inhospitable conditions. They spread rapidly over large areas. Collectively elms are still remarkably catholic in their choice of habitats — surviving quite adequately in groups or in isolation, in coastal or inland regions, at low or high altitudes, on rich or poor soil, and in wet or dry areas alike. The same can be said of very few other trees. By the time the continents achieved their present positions, or something very similar, between five and twenty-five million years ago, elm forests or forests which were mixtures of elms and other trees — such as oak or hickory — were already commonplace features of temperate landscapes throughout the northern hemisphere. By the time man evolved into his specific form (considerably less than a million years ago), several races of elm were weathering a long and continuing series of climatic reversals — ice ages — with conspicuous success. At the best of times, elm populations in general tended to grow in more northern situations in America than in Europe or Asia, a difference which holds true to this day. The vegetation's advance and retreat, in keeping with the advance and retreat of ice caps, has probably always been a smoother procedure in North America, where the major mountain ranges run mainly north to south. In the Old World, the east-west orientation of the Alps, the Atlas, the Caucasus, the Himalayas and so on, has meant that accumulations of residual ice have been liable to delay the reforestation of points north by forming glacial barriers in front of an otherwise welcoming hinterland, when the main boundary of the ice had already retreated back toward the poles.

In about 5000 B.C., when agriculture was rapidly becoming the staple human way of life in Europe and the Middle East, elms were still typically forest trees wherever they occurred.

Though pockets of elm forest still persist in a handful of localities in Europe, Asia and the USA, the habitats of most present-day elms are anything but forests. There are natural reasons for this, but these are insignificant alongside the human factor. The large majority of elm habitats are now human-modified or simply human habitats.

THE ELM IN RECENT HISTORY

The sudden drop in elm populations that occurred in northwestern Europe about five thousand years ago is testified by the notable scarcity of elm pollen traces in lake bed deposits laid down at that time and subsequently. This truly dramatic population decline may have been exaggerated by disease epidemics. But populations of other trees, such as the ash and hornbeam, did fluctuate in the same direction at the same time, although to nothing like the same extent as the elm. This overall trend, then, was quite certainly the result of human activity — following Bronze Age man's discovery that the soil beneath the ubiquitous woodland which then blanketed the European lowlands was many times more fertile than on the hilltop sites men had hitherto favored. By the time historical records began to be made, deforestation had probably laid bare over half Europe's virgin forest tracts. The devastation also coincided with the shift in human society's general preoccupation from hunting and gathering, to herding and cultivating. Increasingly well-organized and politically stable societies took to the moist, forested lowlands that had hitherto been unsettled.

Just as it is now reducing the forest cover of the Third World, shifting cultivation (see p. 32) began to work its rapid way through the primeval forests of Europe. A dozen years was probably the maximum period during which the newly deforested soil could provide a useful harvest. At the same time further woodland was destroyed by deliberate burning to flush out game or to increase the range of grassland for herding. Bigger and bigger herds of domestic animals grazed also in the woodlands and busied themselves with nipping new generations of trees in the bud.

These cumulative practices are thought to have discriminated particularly against the survival of trees like elms, which are thin-barked and therefore have low resistance to fire, and which also tend to rely on root suckers as much as on seeds for their regeneration. Yet all this evidence does not, in my view, account sufficiently for the very extreme reduction in elm numbers that the pollen evidence shows. Besides disease, another possibility that could help to account better for the evidence is that elms may have been felled preferentially by early cultivators because they were more inherently useful as commodities than most other trees.

In these days when western agriculture and commerce are virtually independent of any resources a native vegetation might offer, it is hard to remember that most of the European trees that grow today within a hundred miles of a long-standing center of human settlement are cultivated plants, or are at least direct descendants of trees once cultivated for explicit uses. Oaks, for example, were cultivated up until the late Middle Ages to provide food in the form of acorns. Acorns were then used not so much as animal fodder (as they sometimes still are today) but as human nourishment. All over Europe they were baked, crushed to a powder and used either wholly, or as a supplement to wheat and barleyflour, in making bread. Then in later ages, oaks were frequently cultivated to provide building or ships' timber. Many oaks still standing today in Europe have a more than usually bell-bottomed shape. This indicates that they were pruned in infancy to encourage the growth of their lower boughs, to make them suitable as structural timber in wooden ships.

"Cultivation" in this context meant more than simply keeping a watchful eye on natural stands of useful trees. Perhaps that was the case when agriculture was in its infancy and a seemingly endless supply of all kinds of tree was ready to hand. But by the Middle Ages, when woodland had largely come to be replaced by moorland, grassland and plowlands while human settlement had long since taken the form of permanent villages and homesteads, it makes sense to visualize a situation where men had stranded themselves at some distance from a source of tree products. After many hundreds of years of disuse, large areas of land once impoverished by deforestation and shifting cultivation had recovered a good deal of their fertility. For this and many other reasons medieval agriculture was more efficient, ambitious and productive than its Bronze Age equivalent in every respect. One sure sign of this improved know-how was that trees which were particularly in demand became domesticated plants, and were encouraged — if not compelled — to grow close to houses and home pastures. This domestic model goes a little way toward explaining the relative stability of populations of certain trees — notably oaks and elms — after their initial sharp decline during agriculture's trial-and-error beginnings.

What, then, were the uses to which elms could be put, that might recommend them so highly as harvestable commodities and, later, as homestead trees? We are so accustomed to seeing elms used as shade or shelter trees in towns or around farm buildings and paddocks that we are not at once inclined to question the logic of this arrangement. True, most elms do have certain advantages that fit them for this particular role. They are quick growing, need little soil, are tolerant of pollution and certainly provide a wealth of dense shade in summer.

In other ways, however, they are an odd choice as adjuncts to human settlements. Their enormous shallow root systems are notoriously liable to undermine the foundations of buildings. Their top-heavy winter skeletons are a favorite meal of lightning and strong gales and, even on the stillest summer days, most large species of elm have the disconcerting and antisocial habit of shedding whole boughs without a second's warning on to whatever happens to be beneath. Kipling commented on this trait in his mock-antique poem, "A Tree Song":

> "Ellum she hateth mankind, and waiteth
> Till every gust be laid,
> To drop a limb on the head of him
> That anyway trusts her shade."

Kipling's distaste for elm trees actually proceeded from a sort of arboricultural chauvinism. He had the fixed, but mistaken, idea that elms (unlike the good old oak, ash and hawthorn trees) were not native British trees but were introduced by the Romans. His other observations were, nevertheless, correct in essentials. So the question remains — why encourage a tree like the elm to grow where it could clap several tons of bough like a mallet down onto humans or their buildings?

Elms are not even useful as firewood. Elm logs have to be stacked and dried out for a year or two before they will burn without spluttering and smoking — even then, they burn too quickly to be of much use. Nor were they ever suitable for charcoal. They took too long to cook and the end product was unpredictable in its performance.

The answer to our riddle is probably that the use of elms purely as landscapes or shelter trees is a rather modern development arising, by force of habit, from other, older uses. These were the uses for which elms were originally singled out for felling by primitive subsistence farmers and later singled out for preservation by early settled agriculturalists. It is likely that one of the most important of those uses involved not the timber but the leaves of elm trees. Elm leaves were used, right up to the middle of the nineteenth century in most northern European countries, as fodder for cattle. An elm tree's area of leaves per area of land it shades at noon is around the maximum possible for a large broadleaf tree. The amount of vegetable protein and carbohydrate contained in the foliage of a mature elm in full leaf is equivalent to that in two acres of ungrazed meadow. Elms were "stryped" of this wealth of food in late summer, when other sources of fodder, such as pasture and root-crop leaves, were parched or absent. Cattle evidently appreciated the flavor and consistency of elm leaves and, in the days before cold storage and winter-feeding techniques were available, tree fodder probably figured substantially in the final fattening up of livestock before it was slaughtered *en masse* and salted down for winter consumption.

The actual harvesting of leaves was achieved simply by climbing the tree and hacking out the most accessible parts of the foliage. Elms had the advantage of being easy to grow, easy to climb and robust enough to survive such treatment throughout a long lifetime, given that most of the leaves an elm produces are surplus to the tree's basic photosynthetic requirement.

Other trees, such as the chestnut or the hornbeam, performed a similar role in parts of the world where they are as common as elms have been in northwest Europe. Chestnut leaves are still much-used as cattle fodder in some parts of central France, while hornbeam leaves are favored as fodder crops by farmers living in and near the deciduous woodlands that clothe the Himalayan foothills in northwest India.

The use of trees for fodder during prehistoric periods seems, on the face of it, rather a likely practice. There was no organized management of grasslands other than repeated burning, and the European climate during the Bronze Age was, at times, considerably drier than it is today. Primitive herders might well have been brought to the simple expedient of felling trees to make their foliage available to their animals, when grass was scarce and trees like the elm were superabundant.

Mechanized farming has, of course, made tree stripping uneconomical in most parts of the world where cattle farming is practiced. There is, nevertheless, reason to suppose that elms originally won their place as part of the human landscape on account of this very practical function, and then stayed there subsequently for a multiplicity of additional reasons.

One of the reasons was, of course, the usefulness of elm timber. It is tough, yet easy to cut. Its close, eddying grain makes it disinclined to split. Despite the living tree's susceptibility to fungal attack, weathered elm timber is remarkably rot-resistant even when it is buried and/or waterlogged. These rare and useful attributes have, at different times, naturally led to the use of elm timber in Europe for a rich variety of purposes.

The longest-established application of elmwood has been in the manufacture of coffins. Coffin burials in graveyards became usual in Christian Europe after about A.D. 750. It was during the ensuing centuries that the right of all churchgoers to occupy a grave on church land became enshrined in law in most European states. Partly in emulation of the gentry's grandiose family vaults, partly for hygiene's sake, but also partly to mark out and perpetuate the departed's territorial claim, burial in boxes became the normal funerary practice among ordinary people. Only in very recent times did cremation come to be acknowledged by many as an acceptable and convenient alternative. Even so, cremation nods to the lore of the older practice by continuing to require coffins as part of the ceremonial furniture. So there is no respite for trees here.

Coffins are traditionally made of wood because for one thing the use of other materials could never have been able to keep in step with demand. As late as 1550, only an estimated 30 percent of all females born in England ever attained the age of twenty; a mere 10 percent survived to fifty. The life expectancy of males was probably even shorter. Short lives of course meant more coffins in any given span of time. Until comparatively recently funerals were therefore extremely frequent events in European households at all levels of society. There was a literally endless demand for supply of made-to-measure coffins, a demand that could not possibly have been satisfied by the use of some material less plentiful and less easily worked than wood — stone, for example. But why has elmwood always been such a favored material for grave funiture?

Christian teaching supposes that all flesh is grass, that corruptible mortal remains are inconsequential byproducts of the soul's progress toward heaven. But certain aspects of Christian burial have always harked back to an earlier, superstitious regard for the creature comforts and, in a sense, the preservation of the dead. The use of elm as coffin fabric is probably one such aspect. Elm coffins remain unrotted in the soil for far longer than coffins made of other temperate woods. They represent a compromise with the primitive impulse to build the dead a sound home where they will stay put.

The reality of such thinking is best illustrated by the case of the English landowner who, in the early nineteenth century, wrote a will requiring that his body be encased after death in cast iron and buried in the local churchyard. The church authorities forbade this innovation, not on spiritual grounds but in terms of property law. Churchyards and cemeteries were, they pronounced, consecrated for the eternal use of all God's congregation. To allow any one person's remains to monopolize church soil in perpetuity implied that that person or his family owned the freehold by squatters' rights of a piece of land that properly belonged to everybody — or rather to nobody but God. They might also have added that decay was essential if burial sites were to be reused in future.

It would be tedious to list all the historic uses to which elmwood has been put in the Old World. Worth itemizing are some of the more (to us) recognizable applications, so as to give an idea of the range of demand.

Before ceramic, plastic and metal pipes came to be used to carry domestic water supplies, elmwood conduits (hollowed-out elm trunks) were frequently employed for this purpose. London's water supply during the eighteenth century was channeled mainly in this way. Parts of the system are still unearthed today on construction sites, in perfect condition and working order. This usage brought into play the wet-resistant and durable qualities of the elm. Then, wooden cobblestones, very commonly used for paving in towns far from a cheap source of dressed stones, were normally of elm. Elm timber was also chosen, throughout Europe, for the nonstructural parts of wooden ships, usually in conjunction with oak beams and hulls. In eighteenth- and nineteenth-century England so much of the native stock of usable trees had been exhausted for shipbuilding that landowners were encouraged by financial incentives and some legal pressure to plant oaks and elms for future navies. In times of war, or rumors of war, countrywide "throws" of seaworthy trees were mounted by agents of the crown. The stir caused by their activities was often the first inkling that people in remote rural areas had of a warlike turn in international affairs.

During the canal craze in nineteenth-century northern Europe, the planking of lockgates and flood barriers was usually of elm. When railroads supplanted canals, elm was once more a natural choice for use as ties. Meanwhile, in the booming coalmines of England and Germany, elm pitprops were in constant demand. To this day the mining industry is still a major customer of the elm timber trade.

Aside from these varied and vital technological uses, elm has always been preeminent as a craftsman's material due to its hardness, good looks and easy working qualities. Among other items, tool handles, mallet heads, chair seats, toys, packing cases, bowls and veneer were — and still are — products for which elmwood was the natural choice. No matter how many tropical substitutes reach the market, elm is still the stock-in-trade of thousands of small-scale craft industries.

DED has not affected such industries in the way one might suppose, for there is, surprisingly at first thought, something like a glut of elm timber on present markets. The reason is that the disease does not greatly affect the timber-bearing parts of trees. They can still be used as usual. Nevertheless, because diseased trees are often felled before their prime, or healthy immature trees are harvested in anticipation of disease, the individual mass of trees that reach the open market is diminishing in the long term and the quality of the wood is not as high as it could be. As a result elm timber for fine uses (cabinet making for instance) is nowadays often imported into Europe from Japan and from America, where the elm population is still large enough to provide a steady supply of mature lumber.

What will happen to less specialized elm-using industries when the timber supply runs low, as it certainly will if DED continues to take its current toll? It seems likely that they will by then have expanded their volume of trade by virtue of a plentiful supply of underpriced raw material. In the absence of a home-grown substitute, the industrial gap left by the elm will surely have to be filled by a suitable component of Third World forests. Here we see a direct illustration of how crises in the natural history of trees in the developed world add to the already irresistible strain placed on the survival of tropical ecosystems.

American elms today are generally confined to the eastern states, though they were once fairly plentiful in the west also. Ancient climates are mainly responsible for this present one-sided distribution, but man modification of the environment has also probably played a part. The burning of woodland for purposes of game flushing or cultivation was a routine practice among North American Indian tribes for example. Trees with thick bark, like the phenomenal coastal redwoods, survived this treatment, but any residual elm populations that might have adapted to western climates must soon have resigned from the landscape under such conditions.

When European colonists arrived on the moist eastern seaboard, however, they were gratified to find a huge abundance of elm in natural woodland stands. Colonial place-names like New Amsterdam or New Hampshire were no doubt simple expressions of expatriate loyalty, but it is tempting to speculate that the choice of names might have had something to do with the similarity of the eastern landscape to elm-dominated places in the settlers' native lands.

There have been many economic uses of elm in America. Some correspond to the Old World applications mentioned earlier, others (like the use of elmwood in automobile bodywork fifty years ago) are unique. But the classic employment of the American elm has been as a shade or ornamental tree in towns, particularly in areas originally settled by the Dutch or British. I have suggested that the reasons for this preference are not so much functional or even strictly aesthetic, as historic and sentimental — a nostalgic echoing of European townscapes. Yet the American elm *is* an exceptionally good-looking tree in its own right. The French botanist André Michaux (1746-1802) called it "nature's noblest vegetable" and it is not difficult to see why it found favor as part of the view from American windows. The benefits conferred are not, of course, purely visual. Besides providing dense summer shade with their tightly packed leaf canopies, elms mitigate air pollution by trapping dust and beads of airborne fuel exhaust on their minutely hairy leaf surfaces and smother traffic noise to a measurable degree. They also act as windbreaks, though this competence is probably less important in modern American cities than it had been in, say, the windswept coastal towns of Holland.

All these advantages are arguably economic, in terms of the contribution they make to general productivity and happiness by safeguarding health and taking the edge off the worst strains of urban life. But such considerations were probably far from the minds of the people who first cultivated landscape elms. To see their impulses at the plainest, we can return to eighteenth-century England, where elmscaping had an unprecedented vogue.

Considerable official pressures, as we saw, had long been brought to bear on members of the landed gentry to make them plant trees for military purposes. To these pressures were added the requirements of a series of enclosure acts which, during the eighteenth century, enforced the division of large areas of common grazing land into small fields with distinct boundaries. Though they gave rise to a good deal of social injustice, the acts also led to an enormous increase in the mileage of hedgerows throughout the kingdom. Elms were already common as field weeds and, as noted, as a fodder resource. They took little persuading to become quick-growing, self-extending hedges. The English elm, with its propensity for multiplying by sending multitudes of suckers up from its roots each year soon began to dominate agricultural areas. Plowing or grazing on either side of the hedgerow schooled the suckers into disciplined lines. Brambles and other dwarf thorny plants filled in the spaces with some help from the professional hedger.

The enclosure acts and the potential mobilization of future navies coincided with a craze for landscaping that grew up among wealthy, leisured landowners and gentlemen-farmers during the same period. In many areas, farmland and parkland became blurred into one conscious design. Meadows and fields were planted with strategically placed trees as contributions to elaborate vistas, complete with artificially created hills and dales. Sometimes these prospects were planned with the utmost ingenuity to provide not only visual interest and pleasure, but also an acoustic which could reflect voices raised at particular vantage points through anything up to seven echoes!

That elms were of paramount importance in landscape design in eighteenth-century England is also clear from the work of landscape painters like John Constable. Tourists who nowadays visit the scene of one of Constable's most famous pictures, Flatford Mill, on the Essex-Suffolk border, an hour's drive from London, will find it almost exactly the same now as it was then, except in one particular. The elms Constable depicted towering in the background are no longer there. Nor, since the 1950s, are their offspring, for Flatford Mill stands in one of the worst DED-hit areas in southern England. The difference their absence makes to this familiar view is instantly noticeable and very depressing. It is only fair to add that while elms did stand in the places allotted in Constable's pictures, they were rarely the same elms that the painter portrayed. Constable kept a stock of hundreds of sketches of elms in his sketchbooks and inserted them into his paintings wherever he found the elms on the site of his composition inadequate to his theme. Though (strictly speaking) a deception, this doctoring itself is a testimony to the utility role elms play in the archetypal English landscape. And the doctoring was in any case not confined to the fancy of painters. Planters and landscape gardeners were busy in Constable's time, perfecting new hybrid strains of elm as part of an overall gentleman's agreement to enrich the rural scene. Their efforts benefited from the rapidly growing technology of hybrid cultivation.[2]

The treescaping craze was by no means a phenomenon restricted to England. All over Europe, Scandinavia, Russia, the United States and the Near East much the same impulse was identifiable. In one of her remarkable short stories, the Danish writer Karen Blixen spares a thought for "the unselfish spirit of the eighteenth century — which must have walked between sticks six feet high in order to give coming generations shade and foliage."

That spirit is sadly hard put to find a counterpart today. One might suppose that the advantages of technology would put today's planters way ahead of eighteenth-century amateurs. But that is unhappily not the case. Organizations today concerned with tree planting on a large scale are mostly large commercial concerns or organs of central or regional government. They (whether they wish it or not) have to justify their efforts in terms of cost-effectiveness. Importantly, they have also to compete with agriculture for the use of land. Not so the eighteenth-century landscape enthusiast.

The best illustration of the way times have changed is provided by Alexander Grant, a Scottish Parliamentarian who died in 1714. During the last ten years of a busy life and with the help of a few dozen foresters, Grant planted more trees across his estates and his neighbors' estates than the Forestry Commission — the predominant forestry conservation body in Britain with a personnel of many thousands — has planted anywhere since it was set up about fifty years ago.

The decline of tree planting has much to do with changes in agricultural methods — particularly with the trend toward monoculture in large fields, a system which serves the economics of machine farming. It relates also to changes in land ownership. Country estates (for a variety of reasons) are a thing of the past. Most of them have been parceled up and disposed of long since. Even the countryman farmer is a rare breed, now that farms are increasingly managed as commercial enterprises by hard-nosed, urban businessmen who delegate their fieldwork to hired employees. The shape of the landscape is a matter of small concern to anybody without a personal stake in the appearance and long-term welfare of his land.

All such outwardly practical changes also reflect social changes. Since the Industrial Revolution, the indications and applications of wealth have rapidly changed. The landed gentry have all but disappeared and with them a landscaping ideal that more or less democratic institutions, it seems, cannot properly endorse. In Britain, a law exempting money spent on tree planting from death duties (a tax on inheritable wealth) was recently abolished. Perhaps this was the last knell of the phenomenon of the grand-scale private planter. The change was no doubt made for sensible reasons. But will public initiatives adequately replace the work of leisured enthusiasts? Will the average person find he has extra incentives to plant trees now that the wealthy have fewer? Official planting programs outside national parks are not uncommon and their results are often worthwhile. But so far they are proving, on the whole, much less effective than the efforts of a handful of inspired individuals were in the past.

In America, where landscape planning has not been the exclusive preserve of the rich, economic realities are nevertheless pitted against ideals. National parks and wilderness reserves are all very well, but their whole depersonalized conception is foreign to the true spirit of landscaping, the harmonious mingling of the works of man with those of nature. Where, we can well ask, are the Alexander Grants or the Johnny Appleseeds of today? It is hard to deny that, for all the hard-won technology that the amenity planner now has at his disposal, the future of landscape trees like the elm would look brighter if it were once more in the hands of amateurs.

ELMS IN FOLKLORE AND ART

In most known human civilizations, trees, at one time or another, have been used as religious symbols, or as out-and-out deities. Sir J.G. Frazer's well-known hodgepodge of mythology, *The Golden Bough*, bears witness, from the title page onward, to the potent part trees have played in major cosmologies. The most persistent myth attached to trees is that of a link between earth, heaven and the underworld. Such an image was probably uppermost in the minds of the druids who worshiped sacred groves of elms in parts of western France when Celtic culture had its heyday. A more diffuse brand of mythmaking (such as the Kabbala) plays on trees as images of the growth and branching of individual or cultural destinies. Even in the most agnostic developed societies, trees are still an inevitable part of the furniture of romantic poetry or of town-dweller musings on the joys of rural life.

Elms have had their share of such attentions. But somewhat paradoxically in countries where they are, or have been, common, they appear in literature and art more by implication than by explicit mention. The crucial part elms play in scenes wild or contrived has always been that of a makeweight, a massive blocking in of bare spaces or a screen around eyesores and distractions. These discreet contributions are, again paradoxically, most conspicuous by their absence. DED has, at least, made that point clear.

In European and American folk medicine, "slippery elm" tea was, and sometimes still is, used to treat throat and lung inflammations. The bark of elms like *U. rubra* and *U. glabra* is heavy with a kind of slime, which is the therapeutic principle. The fresh bark, or powder ground from dried bark, is mixed with boiling water, then strained, cooled and flavored with lemon and sugar.

Among the uses to which North American Indians have put the elm, the most important is in the making of tomahawk handles. It is not unlikely that this habit came about not just because elmwood is hard, but also because of its durability when buried.

THE ELM TODAY

Perhaps it is symptomatic of the wispy official attitude to the plight of the elm that nobody anywhere really knows the size of surviving elm populations, or even what the financial value of the elm timber used in commerce amounts to. Annual censuses of elms were mounted by the Forestry Commission in England through the early seventies. But these were estimates based on a scattering of sample areas. However, they indicate a population of between seventeen and nineteen million elms diminishing annually at a rate close to 10 percent.

Admittedly, the logistics of a detailed, full-scale survey of elms would be daunting by any standards, but there are obvious potential sources of help in such an enterprise — schools or farmers' unions for example — which have so far never been canvased to see what they could contribute to make survey projects more feasible.

Even in countries like Germany and Yugoslavia, where the majority of elms are probably concentrated conveniently in mixed forests along the great river valleys, no credible estimate even based on sample populations has yet been published.

In the USA there were probably several hundred million elms still standing in 1960, about half of them domesticated trees. Since disease has so far attacked woodland trees less than urban or farmland trees, the latter are now likely to be in a minority. Where they have been replaced, they have been replaced by other kinds of trees — mainly limes and planes.

The USSR also has a massive elm population but lack of information puts even a chance-your-arm guess at numbers out of the question. In India and other parts of the Far East, elms grow mainly in out-of-the-way highland areas, so the problems of population sampling seem once again prohibitive. And yet without a reliable idea of the present status of elm populations throughout the world, it is impossible for us to begin to monitor the impact that disease may be having on natural systems and human economy.

Most Asian elms, though apparently disease resistant, do not yield usable timber. There are exceptions. "Japanese elm" — a timber-trade term which actually refers to two rather similar woodland species — has high commercial value. Other noteworthy timber trees are rock elm (USA and Canada), American or white elm, English, wych and Dutch elm.

When discussing the future of elm trees, it would be helpful to know the cash value of elm timber country by country. But figures are not available anywhere. Five years ago, the total trading value of elm timber marketed in England was unofficially estimated at £4-6 million a year — a figure which, though modest in relative terms, might surprise the many people who have the impression that the elm is not an economically important tree.

The information vacuum just described also surrounds what is arguably the most vital aspect of the elm's disappearance — its effect on local environment and ecosystems. Cultivated or not, large populations of elms have long figured in well-established ecosystems all over the northern hemisphere. As a result many different animals and plants have come to depend directly or indirectly on the presence of elms for a livelihood. It should therefore be no surprise to find that the ecological side effects of their decline could extend across many habitats, including our own.

To defend this view we have to fall back upon an ecological axiom rather than direct evidence, for no concerted attempt has ever been made to enumerate, let alone evaluate, the elm's relationship to the whole range of its natural dependents and competitors. It would unfortunately take many years of costly research to achieve even a small amount of precise information or insight on this score. So little information exists (outside the concentrated studies made of the ecology of some of the DED-carrying scolytids) that it is also hard to guess even roughly the priority that deserves to be placed on such research.

To give an idea of the kinds of relationships that might repay study, however, a concrete (although still anecdotal) example must be cited. As our example we take the association between European elms and rooks (large, crowlike birds without an exact equivalent in the American fauna). This is a case relatively easy to visualize, though most of the elm's important associates are probably microscopic or otherwise difficult to observe.

Rooks are gregarious birds, nesting in enormous colonies — rookeries — in the tops of tall trees. Often the rookery contains many hundreds of nests and individuals. These birds eat mainly insects and are well known for following the plow while they snatch up worms and insect grubs, including crop pests like the carrot-fly larva. They congregate in fields to forage for soil insects, and sometimes also eat freshly sown seeds, but their influence on agriculture is, on balance, mainly beneficial.

In England, elms, hedgerows and spinneys have always been typical rookery sites. There are certainly many areas where there might seem to be no alternative, that is, where elm is practically a monoculture in its own right. But the nineteenth-century naturalist Richard Jefferies(13) maintained that rooks actually show a strong preference for elms as nesting sites, where a choice exists. The elm's flexibility to wind and the configuration of its upper twigs and branches provide, according to Jefferies, a rook habitat that could not be improved upon. Jefferies also described (in 1879) how elms suffering from disease (a disease incidentally, which might, to judge from the description, have been DED) are abandoned by their rook inhabitants long before any symptoms are apparent from the ground. Once one or two trees have died, the entire rook colony leaves the stand of trees *en masse*, to found a new colony elsewhere.

Modern authorities on rook behavior are skeptical of some of Jefferies's observations and take the view that the distribution of rookeries has little to do with that of elms, yet the current British bird census (a long-term survey of population levels of all British bird species) does show a distinct drop in rook populations throughout the country over the past ten years. Enormous numbers of diseased elms have been felled over the same period, while the incidence of diseased trees still standing may be as high as 60 percent in places. The relative scarcity of rooks is especially marked in agricultural areas badly hit by DED.[3] Though rooks will nest in various other tall trees, there is the further consideration that few trees besides the elm grow around fields in clusters substantial enough to support large rookeries. And again, though rooks will flock across very long distances to foraging sites, if the main centers of rook population have shifted away from DED-struck areas, the birds' contribution toward insect-pest control in those areas will have suffered some reduction. It would be interesting to relate, for instance, figures showing upward trends, region by region, in the use of agricultural pesticides, to a decreasing incidence of elms and rookeries in the same regions. If a correlation were shown to exist between the first two factors, there would be reason to suspect that here was a tangible example of the environmental and economic disadvantages of a wait-and-see policy toward DED. The rooks' population decline could then be considered as one among the many mechanisms bridging cause and effect. Most importantly, proper investigation of these sorts of questions might well make the economics of a concerted DED-control campaign look more reasonable or, dare one say, justified. Though essential to skillful agriculture, pesticides are always expensive and often environmentally hazardous. If their use increases as an indirect result of DED, the consequences may cost us far more than a DED-control campaign would cost in the here and now.

The possibility of a link between elm, rook and insect-pest populations is a hypothesis merely — but it is at least one that could be confirmed or denied by statistical analysis of existing figures. A far more important hypothesis, though less amenable to discussion and testing, is the proposition that the removal of diseased hedgerow elms has had an effect on the perpetuation of hedgerows in general. Such removal, I suggest, is often the cue for the destruction of entire hedges. For, in the absence of the dominant tree component, changed amounts of light, air and moisture reach the soil, and so rapidly alter the inventory of the hedge undergrowth. Thrown out of balance with the crop habitat it formerly defined, the hedge can become a troublesome source of, rather than any barrier to, weeds and pests. Under these circumstances, it may make economic sense to the farmer to replace teeming ecosystems with wire fences. Were the long-term agricultural consequences of hedge removal known in detail, however, the action might seem much less expedient.

In this portrait I have tried to sketch the heritage we stand to lose by the elimination of the elm — and thus other trees — from our fields, towns and woods. Rather against my will, I have also managed to show how remarkably poor our knowledge is of the ecological and economic status of even a familiar tree. It is not the last time we shall be forced to this conclusion in the course of this book. It will become clear as we proceed that, no matter where we turn, the same frustrating dearth of information hampers any serious intention to preserve trees on a meaningful scale.

4 Disease Despite Itself

"One can see . . . how delicately balanced a symbiotic association can be, and how close to parasitism (which has indeed been called 'the most exquisite example of symbiosis') . . . balanced by who knows how many millenia of attack and repulse and final treaty."

ANTHONY HUXLEY
Plant and Planet (12)

THIS IS A technical chapter about Dutch elm disease and similar tree diseases. It describes how such diseases work on a tree, how they are transmitted, and the role of disease in that complex interrelation of plants and animals which makes up the total living world. The nature of tree diseases ties in with the structure of trees. Therefore we also need to look at how trees function and grow.

DED is caused by a parasitic fungus. This fact is nothing special in itself. Most trees are prone to fungal invasions of one kind or another. These invasions are mostly either harmless or actually beneficial. Damaging and fatal invasions are in the minority, no matter how conspicuous or widespread their effects may be to us. Just like any lodger, a parasitic fungus has ultimately nothing to gain and everything to lose by destroying its host, the provider of its food and shelter.

Ceratocystis ulmi — the fungus which causes DED — is basically no exception to this general rule. It is true that *C. ulmi* is unusual in that it is both a parasite and what is called a saprophyte. That is to say, it can flourish both on living and on dead elm trees. Nevertheless, its purpose in life, despite appearances, is not to slaughter elms and remove them from the face of the earth. The death of the tree which *C. ulmi* infests is essentially a side effect.

It is also well to remember that parasitism — the act of living off other living creatures — is the norm on this planet rather than the exception.

Parasitic pests and diseases cause death only by accident, and even then — in relative terms — not often. Parasitic bacteria live inside and around us in their billions. But only about one bacterial cell in every thirty thousand does any harm to man or to plants and animals. This ratio is smaller by almost half than the number of humans who commit murder in the USA every year — around one in seventeen thousand.

Even actually harmful parasites do not remain harmful forever. For if they do, they must end by wiping out their hosts. In so doing they wipe out themselves — unless they chance to mutate swiftly to a more versatile form which can then inhabit a new host.

There is another alternative, which in fact seems to be the rule in nature. It is a tendency toward the stable coexistence of host with pest or disease, given sufficient time and fairly constant background conditions.

There is no lack of examples that show this process at work. The rabbit disease myxomatosis, for instance, had its heyday of destructiveness in Europe and Australasia before the late 1950s when, inexplicably, it seemed to decline in several areas. Investigators found that the parasitic organism which caused the disease occurred in those areas in two forms, one aggressive and one relatively mild. Curiously, when it came to competition between these differing forms, it was clearly the milder virus which gained the upper hand. The precise selection pressures which had brought about the situation remain in part a mystery. But it is surmised, not unreasonably, that the more aggressive strain of myxomatosis killed the rabbits so quickly that there was little time for infected individuals to pass the disease on. The slower-acting form, on the other hand, had ample opportunity to switch to new hosts. Its diseased carrier was able to remain in circulation among other rabbits far longer after infection.

Considerations like these are now regularly applied to the study of disease epidemics. It has been shown that DED is also clearly undergoing rapid changes of its own accord which, on current reckoning, ought to make control of the disease easier.

Unfortunately, however, the mechanisms thought to govern the tendency toward the stability of the host-parasite relationship are not the only ones at work in the case of DED. Other, unnatural factors, of which the interference caused by human trade is only one, combine to overrule the natural process apparently at work in myxomatosis. It is these other factors which truly make DED a disease despite itself.

DED is by no means the only disease which attacks elms. At least a dozen other diseases have been diagnosed with an international range of distribution. This number includes several producing symptoms almost identical with those of DED. Only one, however, regularly kills elm trees on a more than local scale in parts of the USA.

So why, then, does DED itself take the form of a multicontinental killer epidemic when many other, basically similar, diseases do not? To give this question the attention it deserves we must consider some curious details of the lives of trees, of fungi, of insects and of viruses — for all these are potent factors in the DED problem, quite regardless of the nevertheless crucial human aspect.

None of the information is particularly specialized or particularly original. It can all be found in various information pamphlets, technical journals and textbooks. But here these many sources are united into a single account, using plainly defined terms. My aim is to see that no reader of this book goes without the facts to support informed personal estimates of the value of official strategies — strategies supposedly designed to cope with problems like DED.

I have already mentioned that the death of trees through the activities of fungal parasites is, in a sense, accidental. But when deaths do occur, what forms do they take?

They generally follow one of three patterns. Some fungi — including the well-known "bracket" fungi often seen growing like an elaborate yoke around the trunks of woodland trees — kill by weakening the structural support provided by the trunk and branches. The process parallels the way dry rot fungus weakens house timbers, with the same ultimate result.

Other fungal parasites feed directly on the living shoots and roots of the plant until not enough of these parts are left undamaged to keep the tree in working order. Yet others, in the course of actually rather minor feeding activities, manufacture chemicals which as a side effect poison the tree's own supply of food and water.

None of these cases strictly applies to DED. It is true that infection does indeed begin with poisoning. But it is the elm tree itself that is usually the ultimate cause of its own death.

We all know that when humans are invaded by the parasite which causes the common cold, the body produces as a defense a flood of sticky mucus from the membranes lining the nose, throat and lungs. We also know that this defensive reaction often goes too far for comfort, blocking air passages almost faster than sneezes or coughs can unblock them, and also clouding several of our bodily senses.

Elms invaded by the DED fungus *Ceratocystis ulmi* show a fairly similar response, though its effects are incomparably more drastic.

As it feeds on the wood just inside the bark of a live elm, *C. ulmi* excretes digestive substances which happen to be toxic to living cells in the tree. Once the fungus is established in this outermost zone of wood, these substances — and subsequently the fungus' offspring, too — migrate through other parts of the tree in the stem's normal water supply, which travels mainly through the same zone. Certain tree cells react to this situation by producing a watertight gum — sometimes in a free form but mostly as dense blobs and loops called tyloses. These outgrowths now block free passage through adjacent water courses. The likely purpose of the reaction (more precisely called gummosis or tylosis, depending upon the exact form the gum takes) is to block the spread of the fungus and the poison into other parts of the tree by isolating the invasion from the bulk of the moisture supply and its flow.

Tylosis is quite a common feature in tree cell behavior. Many tree species constantly produce tyloses as a matter of course. Others, like the elm, seem to produce tyloses as an all-purpose occasional defense against hostile conditions, including also violent injury and severe drought. In elms, for example, tylosis might conceivably ensure that trees do not lose all their stored water at once during a drought, or during sudden cold spells when water from the environment is unavailable through freezing. And tylosis is, evidently, most elms' stock response to disease.

Unfortunately, this standard defense is of little use against *C. ulmi*. The deadly fungus spreads much more quickly than the gum can forestall it. Soon, the entire stem becomes clogged with thousands of tyloses and even the large vessels in the trunk that represent the tree's largest store of moisture for emergencies (and incidentally for the tree's spring growth) gum up.

In effect, the elm tree strangles itself. Leaves turn yellow, wilt and fall prematurely. Twigs curl into a shepherd's-crook shape. With crown, roots and moisture stores all disconnected from each other, a badly tylosed elm can survive only by chance. In most cases, it is almost impossible to say whether tylosis or direct poisoning finally finishes a tree off. But almost certainly it is the combination of both factors that mainly accounts for DED's tremendously rapid action and its great success as a killer.

DED is technically a vascular wilt disease. In other words, it harms vascular systems, the complexes of cells through which food and water move around inside a plant. When its vascular system is damaged, a plant first wilts and finally dies completely through thirst and starvation. All the trees and most of the plants familiar to us possess vascular systems, and different wilt diseases occur in many different types of plant, including several which are commonly used as food crops.

It is very often the case that an accidental combination of events produces an epidemic. Endemic disease — a low, constant level of infection which attacks only occasional or weak individuals — is a fact of nature. An epidemic occurs when one of these low-level diseases suddenly goes wild. It does so usually through some offbeat combination of parasites, or of parasites and carriers, or of parasites and climate. It is some such previously unheard-of combination which manages to stumble past the elaborate defenses which a given organism, in this case a given plant's vascular system, has evolved through the ages.

As it happens, nowhere are vascular systems more powerful or more vigorously protected than in trees. Trees owe their size and their success among earth's plants as a whole to that special legacy of their evolution.

On the other side, however, nowhere are there greater potential rewards for parasites than in trees. Trees are veritable mountains of life-supporting substances. The tree's vascular system also provides direct routes to any part of the mountain.

DED fungus can and does thrive, and undergoes a population explosion throughout elm populations, only if circumstances contrive to let it "under the skin" of one elm after another. To find out how DED gets through a tree's defenses, we need to know about the structure of those defenses and the function of the inner workings they protect.

First, how do elms work? Like any leafy green plant, they produce their own food by photosynthesis. That is, they build themselves — quite literally — from sunlight, thin air and groundwater, by means of a vital process known as photosynthesis. Briefly, photosynthesis requires the bringing-together of water, carbon dioxide and light in the same green leaf cell.

Carbon dioxide gas occurs throughout the atmosphere; like sunlight, it can usually be had daily wherever it is needed. But most of a tree's water must travel to its leaves from the ground by way of root and stem, for rain and airborne moisture cannot be relied on as a regular source of supply. In a mature elm — not the tallest of trees by half — this can mean that water must climb up to one hundred and fifty feet from the soil to quench the needs of the topmost leaves. One free-standing elm can transpire over one hundred and fifty gallons of groundwater during a single summer's day.

The scale of the tasks carried out by tree vascular systems is remarkable. The proportion of a tree's total water intake that is specifically required for photosynthesis can be as little as 5 percent, but it is a supply so essential to life that it must be guaranteed by a large surplus. This surplus is not, however, redundant, for on its way through the tree it carries dissolved nutrient materials from the soil to parts that can use them. And it fosters a variety of biochemical reactions that involve those materials. The surplus water also helps to keep the tree's temperature even, to maintain the turgidity (skin tightness) of softer cells, to carry photosynthesized food to unlit or nonphotosynthetic living parts of the tree and — not least — to ensure further supplies of water. For just as the wick of an oil lamp draws up oil by burning it off at the top, trees mop up moisture through their roots partly by discharging excess water through their leaves.

The wedge of stem shown in cross section (Figure 5) gives a diagrammatic view across the mainstream of the vascular system of a model plant. We can supplement this view by imagining (rather morbidly) that we are present at the felling of a mature elm. The different layers met with between the surface and the core of the trunk are, incidentally, fairly typical for all trees.

As we follow the saw's progress through an elm stem the first layer encountered is, naturally, the weatherproof outer skin of the bark — its epidermis. This dark, brittle coating is made of a few ranks of dead, empty cells whose surviving walls are rich in lignin — a concrete-like plant substance. This, in woody plant cells, mingles with layers and strands of the cellulose cell wall, to which it is chemically akin. The epidermis is not much more than a weathered and darkly stained version of the next layer of the bark — the cork. Cork is a tree's all-around insulation against cold, heat, flooding, fire, injury, attack by microorganisms and so on. In most elms, the cork is rarely more than an inch or so thick. Other trees, like the cork oak or certain relatives of the balsa tree, can grow a hand's span or more of cork. Wine bottlers, moonrocket designers, almost anybody concerned with products that need very special temperature, moisture, noise or sanitation insulation, know that natural cork has no rival for that purpose. Even the relatively meager cork layer of the elm

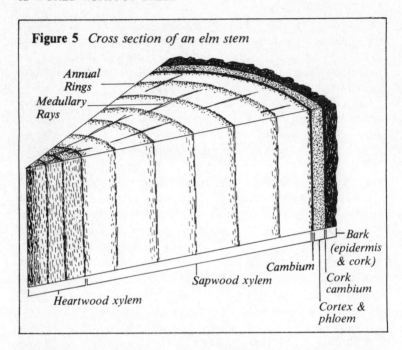

Figure 5 *Cross section of an elm stem*

Annual Rings

Medullary Rays

Heartwood xylem

Sapwood xylem

Cambium

Bark (epidermis & cork)

Cork cambium

Cortex & phloem

makes a vital contribution to its well-being — and, by the way, to the survival of the many small organisms that shelter or nest under the bark of elm logs, and of all trees.

The fissures that run perpendicularly down the bark of elms are mechanically the result of internal stretching, but can also serve to prevent waterlogging in the trunk by channeling off storm water.

On cutting right through the bark into the much paler cortex of the trunk, we should catch a glimpse of a very thin yellowish zone, the cork cambium. Here are the first visible living cells so far encountered. The function of these is to add new cells to the bark as external erosion or wounding reduces it, or internal growth stretches it.

The cortex is yet another mainly dead and protective layer of the trunk, with additional water-holding and weight-carrying duties. The deeper we go into this, the more fibrous and slimy it seems to become as it verges on the phloem (rhyming with "poem"). The phloem transports the products of photosynthesis from the leaves to other parts of the tree. Much of this traffic is in fact destined for the roots, but phloem can and does shift food in every direction. Under a microscope, each of the multitude of vertical tubes that constitutes phloem would appear as a string of long, hollow-looking cells, the connecting end walls peppered with pores. These sieve-tube cells lack detailed contents. The protoplasm, nucleus and most of the organelles have become generalized into a stretchy slime which is slung in continuous strands, like an egg yolk through a sieve, through the connecting pores. Nutrients, produced initially in the leaves by photosynthesis, shunt along within these strands.

Within the phloem layer occurs the cambium, a silk-thin but very alive orbit of growth, which causes the trunk to grow outward — in step with the accumulating weight and thirst of the developing tree-top. The cambium of elms is most active in spring and next in late summer, catering first to the extra water needs involved in the production of new leaves, then to the appearance of flowers and seeds just before the fall. It forms new phloem, cortex and cambium cells at its outer edge and, in its wake, creates xylem, the innermost part of the vascular system and the pipeline for the tree's water supply from the roots.

Xylem is by far the largest component of a mature tree trunk. The other layers can be compared to the rim of a racing-bike wheel. The xylem is the timber-bearing part of the tree and is almost completely composed of dead cells (of ex-cambium).

Xylem cells which the cambium forms in spring have a greater volume than those created in the summer. This difference is easily seen in the annual rings of the wood, so-called because each visible zone of early plus late growth usually corresponds to a year in the tree's life. Tropical rain forest trees, which do not have to synchronize their growth and flowering with seasonal reverses of climate, often do not have this feature. In elms, which are thoroughly distributed in areas of seasonal climate, the rings are always very conspicuous. As I explained, the upward flow of water through the wood is concentrated in the most recently formed rings.

It is easy for water to travel up a very narrow tube, even when pressure from below is minimal. Most xylem cells that actually conduct water are therefore very small in diameter and even the larger xylem vessels that occur in earlywood of trunks and large branches are scarcely roomier — a straw's or a hair's breadth is normal. Many are, however, extremely long. All have minute pits or porous areas in their walls that enable them passively to dump or leak water to the side, into other cells.

A few cells in the xylem zone retain an ordinary boxlike structure with recognizable living contents. These are concentrated in the rays which run, like the spokes of a wheel, out of the xylem and appear to surface through the cambium and far into the cortex as set points. As well as serving storage and lateral transport functions, ray cells often have extra tasks which could loosely be described as medical. They can help absorb and divert concentrations of potentially harmful substances from the xylem flow. And it is mainly from ray cells that the tyloses described earlier originate, snaking into the xylem through cell-wall pits.

After a few years of service, xylem rings and ray cells normally clog up with wastes and mineral residues. They no longer take part in water transactions, but harden to form an extra-solid central support for the trunk. This heartwood can easily be seen in the — by now — completely severed trunk of our imagined tree, for it is much darker and glassier than the surrounding sapwood.

The main working features described above, the vascular tissues of phloem, cambium and xylem, extend throughout the tree from roots to leaves. The veins of leaves contain the same components, narrowed into tight bundles that in turn stem from larger bundles in the twig.

Altogether, a tree's vascular system is elegant, yet marvelously robust in structure. Each leaf is served by faraway roots with hair threads of water whenever necessary. If one channel becomes blocked or damaged, only a tiny part of the treetop will be affected. Many lines of defense protect the total system from injury, unstable weather and disease-causing chemicals or organisms. Elms, which can flourish even in the fouled-up and semidesert conditions of a modern city, evidently have a more-than-averagely tenacious and reliable vascular system.

So how does the DED fungus, *C. ulmi*, manage to infiltrate it with such notorious success?

Since the xylem extends through every part of the tree, it is hardly surprising that a number of different possible access routes are implicated in the success of DED. But *C. ulmi* cannot penetrate either the bark or the growing parts of the elm of its own accord. The possible pathways *C. ulmi* can take are all, except in one case, indirect. The one exceptional case arises from the way elms tend to be found in closed ranks rather than at a distance from one another.

Urban and hedgerow elms are usually planted together in avenues. Underground, their root systems spread over a large area and inevitably interlock with those of their neighbors. In many cases, the roots of neighboring trees actually join forces and become grafted together. But even unmanaged or natural populations of elms commonly share roots. For they, too, tend to grow close together and, as we have seen, frequently multiply not by seeds but by suckers.

In either case — in manmade or natural surroundings — a healthy elm can easily become infected with DED via the roots of a diseased neighbor. The fungus simply crosses from the xylem of the sick tree's roots into that of the healthy tree. Thence it is drawn up into the trunk and branches, where it very soon multiplies. Often it will kill the tree within a single summer.

This type of infection used to be thought rare. Nowadays it is known to play a major part in DED epidemics. In many parts of the world (including the eastern USA) it accounts for as much as half the number of DED victims recorded each year.

Where the roots of neighboring trees are not joined, it seems plausible that certain soil-dwelling organisms — mites, for example — could accidentally carry *C. ulmi* across the gap between sick and healthy trees, to enter the latter's roots through any small wound they may have recently suffered. Above ground, too, there are tree-dwelling animals, especially squirrels and woodpeckers, which could play a similar part, wounding the bark of healthy trees while carrying the sticky, fungal material from a sick tree on their mouthparts. There are in fact no cases on record of such natural connections. But tree surgeons and woodsmen have, on several known occasions, spread DED by failing to sterilize their pruning or lopping tools between operations.

The classic route *C. ulmi* actually takes into individual elms and, more importantly, between widely separated elm populations, is the route provided by bark beetles. It is basically the connection between *C. ulmi* and these insects that has given DED its present significance and which continues to frustrate the efforts of all technologists devoted to the search for a solution to the problem.

The beetles which carry DED belong to the worldwide family *Scolytidae*. This zoological grouping is about two thousand species strong. Its members are together known informally as scolytid beetles or scolytids.

What all scolytids have in common, apart from the anatomical features that show they are related to one another, is a habit of spending nearly the whole of their lives in galleries excavated either beneath the bark or actually deep in the wood, roots or cones of trees and woody shrubs. They emerge from the darkness only for a very brief period of flight which enables them to mate and find new breeding places in another host tree or log.

Most scolytid species show distinct preferences for a particular tree species, or for a selection of just a few tree species. Different, harmful scolytids may concentrate their activities on healthy, living hosts — in which case they are termed "primary" pest insects — or on dead or sickly trees as "secondary" pests. In terms of numbers of species, the majority of scolytids are harmless or secondary pests.

Regardless of their actual pest status, however, many scolytids specialize in invading particular parts of host trees — perhaps the roots, or especially thin or thick portions of the bark. Under "outbreak" conditions (see below) a fairly harmless secondary pest can suddenly become a serious primary pest. A continuing disease epidemic is often the result.

The causes of outbreaks are difficult to generalize about, but a very common prelude to outbreaks involving scolytids is a sudden increase in the numbers of dead or invalid host trees available for exploitation. Such an increase may, for instance, rise after a drought or a year of high gales. In this situation, ever-larger beetle populations accumulate in, and later emerge from, the extra breeding facilities. Pressure of numbers impels the "bulge" generations of beetles to invade living hosts or living parts of hosts they would not normally favor. In many cases the novel host sites are in the vital growing or water-conducting parts of the plant. The burrowing activities of the beetles themselves disable the host tree to the extent that it dies back, or dies outright, making still more weak or dead tissues available for use as breeding material. The vicious circle of outbreak goes on turning until outside influence puts a stop to it. Such influence may take the form of, for instance, a period of unusually cold weather leading to a high mortality in the beetle population, so reasserting the natural balance. Or it might be the ultimate lack of exploitable hosts. The mechanisms that operate at the beginning and end of pest outbreaks are, in fact, more complicated than it is possible to suggest within the scope of this book, and moreover, the ecological studies relevant to understanding such mechanisms are still in their infancy. The basic cycle is simple and plausible enough, but it is subject to endless variations. The involvement of fungi in scolytid pest outbreaks provides a small example of the "wheels-within-wheels" complexity we are dealing with in these fields.

Though many scolytids cannot of themselves become serious primary pests, they often achieve the same effect in combination with a fungus. This is particularly true of "wood-boring" as distinct from "bark-dwelling" scolytids. The forays of the former type into healthy wood, though not necessarily doing much serious harm in themselves, provide an entrance which can be exploited by wood-digesting fungi — fungi which are otherwise not equipped by themselves to penetrate the tree's epidermis.

The microscopic spores (in effect the seeds) of some fungi can arrive unaided at the wounds caused by scolytids in the host tree; they are windborne in invisible multitudes and some are always liable to chance on the wounds, lodge in them and germinate.

Certain fungi — especially those known as "ambrosia" fungi — have evolved ways to exploit the cooperation of beetles, leaving much less to chance. Their actual presence in a piece of wood makes the wood physically and chemically attractive to certain wood-boring beetles. Like the brewers' yeasts to which they are related, ambrosia fungi can predigest their food and cause it to ferment. The fermented products are in fact more nutritious than the raw material and more appetizing to "ambrosia beetles" — the beetles (scolytids and others) which associate with ambrosia fungi. During their flight period, the beetles may first of all disperse haphazardly in search of their preferred host tree species, but they are eventually drawn in great numbers to individual host trees that have been acted on by an ambrosia fungus. The flights are guided by minute traces of chemical attractant borne on the wind, and originating either from the rotting wood itself, or from the excited glandular response of other "pioneer" beetles that have happened across an existing fungal site.

Wood-boring scolytids excavate their breeding galleries deeply into these sites, and the offspring that later hatch from the eggs feed on a mixture of wood and fungus at all stages of their development. When they become adults and the right temperature conditions eventually prompt them, in turn, to eat their way out of their birthplace and search for new wood, they fly off, carrying in their gut and on their bristly bodies large supplies of fungus and fungal spores. Whenever they now land and feed on a host tree they are liable to inject it with a fungal infection. So the foundation is laid for yet another breeding site.

The relationship between ambrosia beetles and ambrosia fungi is so ancient, and evidently has such considerable survival value, that some ambrosia beetles actually sport pits and pouches on their heads and bodies. These have evolved over millions of years expressly as vessels in which the privileged fungus can be transported.

The ambrosia connection can pose terrific pest problems, especially in the tropics where farmers and planters can usually least afford to cope with them. One of the best-known examples is the "shot-hole-borer" beetle's relationship with an ambrosia fungus which annually decimates the tea harvest in Sri Lanka. However, there is hardly a plantation crop in the tropics that is not threatened by similar alliances between carrier and fungus.

Though they do not possess features as elaborate as fungus glands, some bark beetles can also become associated with fungi in much the same overall fashion as wood-boring ambrosia scolytids. The elm-bark beetles which spread DED fall into this category.

All but one of the four known major DED-carriers belong to the beetle genus *Scolytus*. The odd-one-out, though also a scolytid, is *Hylurgopinus rufipes*, a species thinly distributed in the eastern USA and neighboring areas of Canada.

The beetle which gives rise to most DED damage both in America and Europe is the small European elm-bark beetle, *Scolytus multistriatus*. This species was first introduced into America between 1903 and 1909, but it did not act as a DED-carrier there until the early thirties, when its representatives were repeatedly shipped into the USA in cargoes of fungus-infected elm logs. Thence *S. multistriatus* spread from railheads in and near Ohio and New York until, by 1940, DED was in evidence over about eleven thousand square miles of New York, New Jersey, New Hampshire and Connecticut. The distribution of *S. multistriatus* itself by then went considerably beyond those areas, but the disease had already contributed to the death by felling of about four and a quarter million elms, leaving about one in eight thousand elms in the epidemic area diseased, but still standing.

Today, the beetle *S. multistriatus* can be found across most of Ontario, Quebec and southern Manitoba, and in fact in all but three of the mainland states of the USA. The disease it has brought in its wake now occurs in all but ten states. Exact count of the consequence in terms of elm victims has been lost, but the yearly toll of dead or dying trees in the USA is probably well above the estimated world average of one in eight.

As Figure 6 shows, *S. multistriatus* has an equally vast distribution in Europe and the USSR. As always, DED is never far behind it. In some outlying areas of Europe (and in Britain especially), *S. multistriatus* shares its role as DED-carrier with the larger, but closely related, *Scolytus scolytus*. In parts of Scandinavia, *Scolytus laevis* is considered to be the main culprit.

It is generally supposed that none of the four scolytids mentioned

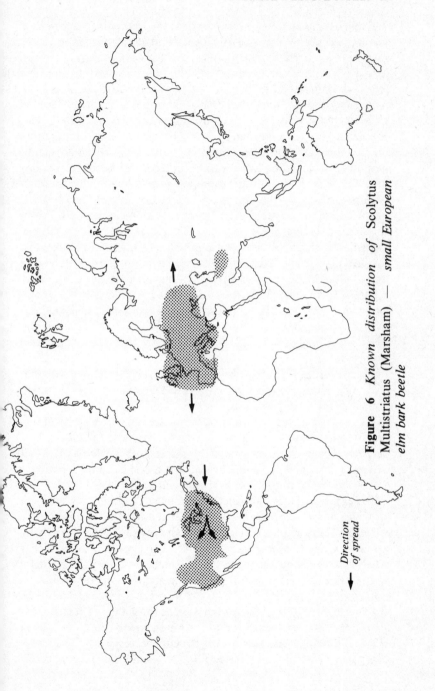

Figure 6 *Known distribution of Scolytus Multistriatus (Marsham) — small European elm bark beetle*

Direction
of spread

had much to do with elm disease until the present century. *Hylurgopi-nus rufipes*, the native American species, was previously known as a rather indiscriminate feeder on the trunk bark of a variety of trees. It is presumed to have picked up the habit of feeding on fungus-rotted elmwood only since 1930. In Europe, the three *Scolytus* species were, with good reason, considered rather minor pests until the arrival of *C. ulmi*, supposedly from Asia, during the present century. Indeed beetles and fungus alike could sensibly be regarded (in isolation from one another) as beneficial organisms, in the sense that they can be numbered among the many creatures that contribute to the breaking down of dead or sickly vegetation into soil-enriching nitrates. Elm-bark beetle attack, unaccompanied by *C. ulmi* attack, incidentally, only leads to the death of elms when feeding or nest excavations are so plentiful as to girdle a treetrunk. Then they interrupt the flow of nutrients through the entire circumference of the phloem zone, and so starve the roots. Such circumstances would only normally arise under outbreak conditions. In earlier times an abundance of natural predators probably also helped keep beetle populations below out-break or crisis level. But modern farming conditions and methods can fairly safely be said to have restricted the activities of such preda-tors — particularly in the case of insect-feeding woodland and hedgerow birds.

The life cycle of elm-bark beetles and the way it interacts with that of *C. ulmi* is best shown with reference to the most destructive of all DED-carriers, *S. multistriatus*.

Like most beetles, *S. multistriatus* passes through a series of dis-tinct developmental phases on its way to maturity. Like many bee-tles, it hatches from an egg as a legless grub or larva. Then it undergoes a complete change to become a buglike pupa. Finally it emerges from its pupal skin as a winged adult.

The eggs are laid by the parent beetle in a straight groove, which it plows in a root-crown direction under the bark of an elm log or unhealthy elm branch. The groove also disturbs the underlying wood to a depth of a millimeter or so. The eggs are laid in niches evenly spaced along both sides of this maternal gallery. The precise number of eggs varies with the size of the gallery and other factors, connected with the fitness of the parent and the presense or absense of favorable conditions inside and outside the gallery. These include the rottenness of the wood and the degree to which the log or branch as a whole is dried out by exposure to sunlight. A typical *S. multistri-atus* gallery extends for three to four inches and houses about two dozen eggs. Some are much longer and can contain eighty eggs or more. Records actually exist of more than two thousand beetles growing to maturity from the eggs contained in one square foot of bark.

If the eggs are laid just before winter comes, the parent may herself remain in the gallery and die there. In a few cases she may survive until the following spring. However, if the parent lays her eggs in the summertime, she may simply fly off to repeat this operation several times in other locations.

Winter-hibernating generations of beetles naturally take longer to develop from the egg than summertime generations. During warm summer weather the larvae hatch from the egg after an incubation period of about a week. They immediately begin to tunnel at right angles to the maternal gallery, against the grain of the wood, creating the rather decorative herringbone appearance we see in grub-infested wood. The tunneling continues for four to five weeks. Meanwhile the larvae feed voraciously as they go. If the yeasty-looking fungal mass of *C. ulmi* is present (as is often the case), this, too, finds its way into the beetle grub's maw. Probably, by causing fermentation in its gut, this actually aids the insect's nourishment. The presence of the fungus may also raise the temperature of the gallery slightly and so help the larvae to survive in cooler weather. After a month or two of tunneling and feeding, the grub enlarges the end of its byway into a pupal cell. There it becomes a pupa, inert but displaying rudimentary adult features. In the course of approximately a further two weeks the pupa undergoes the final metamorphosis into a winged beetle.

At this point feeding is resumed and the now adult beetle may become gorged and smeared with sticky fungal material. If outside conditions are warm enough, the beetle and thousands of its like-minded contemporaries eventually chew their way into the light of day. From outside, their exit holes resemble the holes made by shotgun pellets. Though pathologically light-shunning up till now, the adult beetle is suddenly stimulated by the appearance of light into a burst of flight and mating activity. Before mating, however, newly emerged beetles must feed further before they become sexually mature. It is this maturation feeding that is crucially responsible for the spread of DED — for many beetles choose to perform it in young, living parts of healthy elms. These sites are not qualified as breeding grounds but are nonetheless richly nutritious. *S. multistriatus* adults emerging in the spring commonly feed to maturity in the easily ruptured junctions of young twigs high in the crowns of healthy elms. During the warm spring weather, hundreds of thousands of feeding beetles may congregate in the crown of a single tree. Each eats a deep notch into the V between two twigs, piercing the vascular tissues. Fungal livestock, usually in the form of spores, may now become smeared into the notch by the carrier beetles jockeying for mouth-holds on the sap-rich wood.

In springtime, the specially large earlywood cells of the current annual ring have been newly formed from the elm's cambium and have just lost their end walls and most of their contents. Dissolved cell walls and protoplasm add to the attractions of the sap now flowing thickly toward the spring leaves. Minute fungal spores enter the xylem of this or previous season's earlywood. They travel easily along the transpiration flow, quickly reaching the bases of the new leaves. They also travel, though at a much slower rate, against the direction of the xylem flow. Wherever they become wedged in the xylem's ribbing, they germinate to form one of the fungus's feeding stages, a vegetative mat of fine strands (hyphae) intertwined to form coarser strands, called mycelia. These strands, in turn, are capable of passing through further developmental stages until they produce further generations of spores. In due course, the vegetative phase may also grow across the grain of the wood, passing through cell-wall pits in the dense latewood of other annual rings to gain complete access to all functional parts of the sapwood.

Initially, however, the spread of the fungus is confined to the vertical colonization of a single annual ring. Whenever fungus reaches a major junction of boughs, the tree's foliage begins to die from that point outward, both from poisoning and tylosis effects. Infected areas can easily be seen in a sick tree during the warm part of the year. They are yellow, brown or completely bare against the green of unaffected boughs. If bark is now stripped off the dying branch, the progress of infection can often be seen as a series of brown streaks along the length of the limb. When infection is limited to an internal ring of xylem, a cross section cut across the limb shows a circle of brown smudges.

Now, if ever, is the time to halt the progress of the disease. The most effective — and the safest — DED treatment currently available is fungistatic injection. A chemical which arrests the development of *C. ulmi* is annually squirted into the xylem flow of the elm, near the root collar at the base of the trunk, at several points. The equipment and labor required for this technique can be expensive. If only a few elms are to be treated, doses of fungistatic chemical are in the form of individual packs which can be pegged into the treetrunk by syringe-like extensions — acting much like a hospital drip.

If, however, the tree is given no treatment, the diseased limb (for reasons already given) will now attract beetles seeking a feeding site. They bring fresh batches of fungus acquired either in their natal galleries or from contact with fungus-carrying sexual partners. So the tree suffers wave after wave of infection — an onslaught it can hardly be expected to resist. It has become quite commonplace for growing elms in areas badly hit by DED to die only a matter of weeks after infection.

A dead elm, unless it is removed or debarked, continues to serve as a meal and as a breeding pool for both fungi and beetles for about two years. After this period the sapwood becomes unpalatable, presumably owing to the action of microbes.

The disease cycle just described is subject to a certain amount of variation. In areas of favorable climate, it may be repeated several times during the course of the year. Trees infected during the late summer, however, seem to resist the disease slightly better than in the spring, probably because the summer latewood, as mentioned, is stringier and less penetrable. The time taken by beetles to reach maturity is, nevertheless, considerably shorter in the summer than in the spring.

There is a certain amount of variation in the susceptibility of different elm species to DED. Susceptibility in the three species most often featured in the scientific literature, namely the Dutch, English and American elms, increases in that order, but all are highly susceptible by comparison with some almost-immune Asian elms, such as *U. pumila*. Anatomical features, like the length and breadth of wood cells and the degree to which they are interconnected, appear to be operative in bestowing either susceptibility or immunity on different elms. Other factors, such as the nature of the soil in which different species grow or the chemical composition of fluids present in different elm woods (some of which may have a repellent effect), have also been suggested as determinants of susceptibility, but no clear pattern has yet emerged from research into such factors. While still on the topic of variability in DED's causes and effects, we should not forget that the behavior of *H. rufipes*, the native American elm-bark beetle, differs from that of its European counterparts in at least one important respect. *H. rufipes* is liable to perform its maturation feeding in the trunk, rather than in the treetop of elms. It is therefore potentially an even more insidious polluter of elms than is *S. multistriatus*, since trunk infection is normally the final presage of death in diseased trees. The competitive presence of *S. multistriatus* in *H. rufipes's* native haunts seems to have smothered this potential. But it does not automatically follow that *H. rufipes* could not become a more serious carrier if it were accidentally introduced into Europe. The European climate would support it and the selection pressures which oblige it to play second fiddle to *S. multistriatus* in the USA may be absent in Europe. Moreover, it has already made the journey across the Atlantic more than once. But *H. rufipes*-infested timber cargoes have, luckily, been intercepted by port inspectors or, if they have escaped inspection, have contained only dead beetles. So very many inscrutable factors are involved in DED, as in all disease problems, that any shift, however slight, in circumstances is liable to set the battle to save elms even further back than where it started.

Nothing illustrates my point better than the recent history of DED in Britain. After destroying millions of elms during the 1930s, the disease underwent a curious lull, and retreated into a few remote areas. Even in those areas, it seemed to have become mild in its effect and was almost manageable. Infected trees usually stood a good chance of recovery after a season's dieback. In the late 1960s, however, Britain experienced new DED epidemics of quite unprecedented proportions. This explosion continues, to this day, to impoverish the elm-dominated landscape of lowland England, altering its appearance almost beyond recognition.

Careful research into the reasons behind this resurgence has uncovered the fact that DED (like myxomatosis) exists in at least two forms of differing aggressiveness. Ironically enough, considering that the disease was first introduced into America in insanitary European elm timber destined for veneer plants, it turned out that the new, more aggressive form of DED that spread across England after 1967 had probably been reintroduced from America in insanitary American elm timber destined for veneer plants. The conclusion of most researchers was that the aggressive form had been responsible for the British epidemics of the thirties but had since undergone a mutation into the milder form in parts of Europe, while continuing to flourish in its original form on the other side of the Atlantic. The reintroduction of the aggressive form had not only produced its own brand of destruction, but had also reactivated the spread of the milder form by making more dead elms available for exploitation by beetles.

These events prompt us to consider the historical evolution of DED. Can we find there an explanation of its mutant characteristics?

Most historical accounts attempt to place the origins of DED in Asia. However, this widely accepted view seems to me to be based on some very dubious reasoning.

The existence of a high degree of DED resistance among most Asian elm species has, for example, led several commentators to the conclusion that prehistoric epidemics of the disease had brought about the extinction of all but the hardiest Far Eastern elm species. The fungus (and presumably its prehistoric carrier) is then supposed to have retreated into obscurity, feeding only on dead or wounded trees. In the late nineteenth century, however, the fungus is thought to have somehow jumped into Western Europe and into the range of distribution of European elm-bark beetles, until a regular interaction between the alien and native organisms became established over the course of a few decades.

Several facts undermine these assumptions. First, as Figure 4 shows, the majority of fossil records of species (mostly extinct species) of elm occur in Western Europe and America, although the absence of Asian fossil records can be accounted for by their not having been searched for thoroughly enough, or by conditions unfavorable to the laying down of fossils during the crucial period.

Resistance to DED is certainly remarkable in most Asian elms. Yet there do exist several species of elm in the Far East which grow in the same regions as resistant species but are themselves only semiresistant. In fact, they are in some cases more susceptible than certain semiresistant western elms. How, incidentally, did these western trees acquire resistance? By coincidence? As the result of pressures quite other than disease, or at least other than DED? If so, why should not the same be true of Asian elms?

It is the notion that *C. ulmi* jumped, as it were, from Rangoon to Rotterdam in one fell swoop that stretches belief to the limit. How? Not as a stowaway in timber, for the small Asian elms are ineligible as commercial timber by virtue of much the same qualities that are thought to account for their disease resistance — the fibrous, easily splintered nature of the wood. Perhaps, then, a keen nineteenth-century botanist imported the fungus on Asian elm specimens for an ornamental garden? Why, though, should he bother to bring saplings when he could bring seeds? Or, if he brought saplings, why would he choose the wounded specimens in which alone, by all accounts, *C. ulmi* could settle?

Or perhaps, then, there was no jump, but instead a continuous east-west spread of disease which gradually linked up with a west-east spread of carrier beetles.

But the easternmost records of DED in Europe, the USSR and Turkey have been shown to refer to infections which clearly came from the West (not from the East), usually by sea. If DED traveled from Asia westward into the sunset, why did it not stop before it reached the Netherlands? And why, more to the point, did it leave no trace of itself in passing through? *C. ulmi* has never, after all, been found east of Iran except for one isolated record in Kashmir, hemmed in by mountains and deserts. This latter record is certainly intriguing, but it hardly passes as evidence for DED's Asian origin. The chances are that it, too, is a freak introduction from the West.

Where then did DED come from, if not from Asia? I imagine the question will probably never be answered once and for all. But one possible and quite novel explanation for the disease's sudden appearance in Europe arises rather disconcertingly from recent research into *C. ulmi's* "double identity." If the DED fungus can, apparently, mutate from aggressive into less-aggressive forms, could not the reverse mutation have possibly taken place in the past, begetting infective strains of the fungus from stock so nonaggressive as to have escaped attention entirely until the present century? In other words, could the relationship between *C. ulmi* and European elm-bark beetles perhaps not have been so recently formed as history suggests? Could that relationship have prospered in Europe's remote past and subsequently lapsed, to be renewed during the present century?

Tree rings and other, less concrete documentary evidence do suggest that something very like DED existed in Europe before the mid-nineteenth century. Remember, besides, that it is only since about that time that the life sciences have become sophisticated enough to recognize the cryptic biological sources of disease in humans, let alone in plants.

Elms were four or five times more plentiful in Bronze Age Europe than they are today. They then appear to have ceased to be common forest trees over the course of a mere five hundred years or so. Botanists and archaeologists often ascribe this dramatic population crash to the direct and indirect impact of neolithic man on the forests which, until the end of the Bronze Age, enshrouded almost every part of Europe in near-continuous shade.

Do Bronze Age farming and grazing add up to an adequate explanation for the elm's colossal decline in Europe? Even modern man would find it difficult to accomplish destruction so thorough. Besides, trees like the ash clearly recovered well from the impact of early farming; and other trees liable to suffer the same fate (the hornbeam, for instance) also seem to have managed to retain their identity as woodland plants.

I should like to suggest that a surge in human farming activities was only partly responsible for the elm's decline, and that the enormous quantities of decaying elmwood that slash-and-burn cultivation brought about may have led to outbreaks of pest activity very similar to the epidemics of DED we are witnessing today. Elms felled to make their foliage available to browsing cattle, elms felled in clearing activities but difficult to burn, elms half-buried but slow to rot: precisely the conditions liable to give rise to an outbreak of insects which breed in dead elmwood. The reemergence of the elm as predominantly a nonwoodland tree suggests, to my mind, that the elm left the forests in response to natural, as much as to human, pressures. Indeed, we have seen that the elm's subsequent success as an open-country tree was to some extent itself encouraged by humans.

If DED did exist in Europe in antiquity as I have surmised, it must follow that it subsequently fell dormant, apart from minor flare-ups that left no impact on the fossil record, until the present century. What mechanism can therefore have provoked the drastic alteration in DED's virulence after 1850 that this view supposes?

Consider, first, the summary of DED's recent official history.

1917-18: Symptoms of a hitherto unknown elm disease are recognized in Holland, Belgium and France. Its exact cause is unknown.

1919-25: A spate of research in the Netherlands following the destruction of most of Rotterdam's elm trees provides the first proper

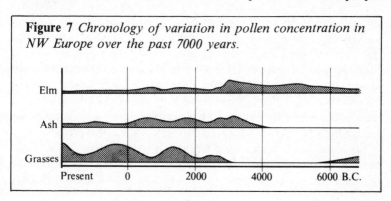

Figure 7 *Chronology of variation in pollen concentration in NW Europe over the past 7000 years.*

knowledge of the disease mechanism, and gives rise to DED's common name.

1926: DED is recorded in Britain, near London. By 1931 epidemics of the disease are rife in southern and eastern England. Over 50 percent of the hedgerow and woodland elms of the East Anglian region have abruptly been destroyed, giving a prairielike atmosphere to farmlands in the area and intensifying soil erosion problems there.

1930-33: DED is accidentally introduced into America and joins forces with native elm-bark beetles and with *S. multistriatus* (introduced twenty to thirty years earlier).

1940-67: Though showing a continuous, rapid increase in North America, DED appears to fluctuate in its effects back in Europe. The introduction of the disease into new areas (Sweden, Russia, Germany and elsewhere) is frequently noted during the 1950s. But in areas where DED epidemics have been long established, the onslaught often becomes noticeably less intense. An epidemic of DED near Vienna, for example, comes to a complete halt of its own accord, while in Britain the disease has become almost a rarity. Meanwhile, in the USA, a national campaign to control the rampaging spread of DED by spraying trees with insecticide is abandoned in the sixties for environmental reasons.

1967: DED flares up anew in Britain. UK Forestry Commission scientists identify two forms of the fungus at work, a "fluffy" highly aggressive strain and a "waxy" slower-growing, mild strain (so named for their outward appearance and behavior when cultured in the laboratory). In areas where both forms occur, the waxy strain is often found to be more widespread than its more virulent counterpart. Also, fluffy *C. ulmi* is found, on several occasions, to transform into waxy *C. ulmi* when kept in a laboratory for long periods. In the USA, however, the fluffy form remains all-powerful: not that the difference matters much, since even the very small populations of waxy fungus that occur there prove to be well able to kill American elm trees. The elm now ceases to be a premier shade tree in the USA except in the few cities where active control measures are adopted.

European researchers report highly variable ratios of waxy to fluffy *C. ulmi* populations in different Old World regions. Fifty years of patient research (mainly in the Netherlands, the USA and Britain) to breed an elm resistant to DED which could be used to replace lost trees goes back to the drawing board — most of the cultivars that had been bred prove resistant to the waxy, but not to the fluffy *C. ulmi*.

1973-75: In Britain, especially in the west and the Midlands, the elm population continues to fall at the rate of about three million a year. In 1975 only about seventeen million elms remain of a prewar nationwide population at least twice as large. In some English counties, nearly 60 percent of elms that still stand are infected with DED. In 1974 (European Conservation Year) legislation is finally introduced to prevent the import of insanitary timber into the UK — but the laws are relaxed a year later to admit EEC imports. No safeguards are placed on the reexport of elm timber of non-European origin within the EEC. Despite pressure from forestry and conservation groups, hardly any statutory limit is placed on the transport of dead elms within the national boundaries of any country where DED now exists.

The reason for setting out DED's history in this way is not primarily to raise questions of control at this point but to stress the bewildering way in which uncontrollable disease epidemics can annihilate sections of the vegetation of entire continents, then, in some places at least, suddenly die out as conspicuously as they sprang up. Things do not always turn out that way. Chestnut blight took only a few decades virtually to exterminate chestnut trees in the USA and showed very few changes of pace in the process. Chestnut blight was a clear case of an introduced alien fungus (this one really did come from Asia) set free from its natural enemies and not particularly dependent on the assistance of insect or other carriers. DED in the USA, though very much dependent upon movements of bark beetles for its spread, is otherwise on a par with chestnut blight in its destructiveness. But DED in Europe is more variable in its effects. Here we have yet another indication that DED may not be an introduction there at all, but a longtime resident.

It may seem paradoxical to argue simultaneously that human interference can be held responsible for the dissemination in the West of DED on the one hand, while urging the possibility that DED may always have been an incipient natural foe of some western elms on the other. Yet these views can be reconciled by reference to a single phenomenon — the same phenomenon which probably lies behind the existence of separate, mutant strains of *C. ulmi*.

A good deal of research into *C. ulmi* mutations is now proceeding on the assumption that they may be caused by a virus infection — that is, by a rival parasite invading a fungal parasite (*C. ulmi*) in classic fulfillment of Swift's adage:

> ". . . a flea
> Hath smaller fleas that on him prey;
> And these have smaller fleas to bite 'em,
> And so proceed *ad infinitum*."

The milder, waxy *C. ulmi* is thought, by most observers, to be the parasitized form. The fluffy, aggressive strain is, by implication, the more healthy or normal. Waxy *C. ulmi's* relative success in areas where both forms are present may arise from its greater ability to operate in harmony with the lifestyle of its beetle associates. Fluffy *C. ulmi's* apparent selective advantage (its superior ability to kill elms) may be no advantage at all. Hans Heybroek (a leading expert on DED in the Netherlands) recently made the point (9):

> It does not seem necessary for the survival of the fungus for it to kill elms: it seems able to survive and complete its cycle just on beetles and in their galleries. Maybe this is what it did in Britain up to 1967, just causing some slight disease symptoms in an elm here and there as a by-product. It could be hypothesized that during such a prolonged saprophytic stage the fungus would tend to lose its virulence just as it does when we keep it in culture for years instead of inoculating it into elms every now and then. It is also conceivable that when both a more aggressive and a less aggressive strain are present, the latter may have an advantage over the former, in being better adapted to this type of life. . . . For the fungus, survival lies above all in riding a beetle or growing in a beetle's gallery. No matter how successful the fungus is in killing a tree, its parasitic phase in the interior of a tree is a blind alley for the fungus unless it is picked up again at some phase by a beetle. . . .

The argument seems plausible, but it does not quite fit all the facts. Why, for example, does the fluffy, aggressive *C. ulmi* predominate in the USA, where native elms are highly susceptible to DED, while in Europe, where elms are more resistant, it often yields its place to the milder strain? If ability to kill elms is a relative disadvantage, things should happen the other way around.

Besides, the assumption that the presence of a viral infection in *C.
ulmi* should necessarily result in a weakening of the fungus's vigor is
not necessarily consistent with the effects viruses are known to have
on other organisms. In other words, the contrary position could
quite possibly be the true one. So the aggressive strain, not the mild
strain, may be the upshot of parasitism. Furthermore, even if aggressiveness is currently the norm of behavior in *C. ulmi*, it is still possible that that norm may have been set after mutation of *C. ulmi* from
a strain normally so mild as to have existed in Europe in almost complete obscurity for thousands of years.

I should add that any firm assumption about the action of viruses
on *C. ulmi* is, in any case, premature. For no specimens of virus have
hitherto been discovered in samples of either strain of the fungus.
The likelihood that one will be found does, however, seem strong in
the light of previous knowledge of the way viruses operate. Such
knowledge is of course itself extremely recently acquired. Most of
the discoveries from which it stems have been made only during the
last thirty years. In that short time, however, ample evidence has
been found for the existence of direct links between the behavior of
viruses and the circumstance of genetic mutation in several creatures.

Though often lumped together with bacteria and other very small
entities under the vague term "microbes," viruses are in several
respects unique among living creatures. So unique that some observers used to declare that the word "living" was inappropriate to
them. Their greatest peculiarity lies in the fact that they cannot reproduce of their own accord but must invade a living cell and subvert
some of its reproductive facilities in the interest of the production of
new virus particles. Viruses are mainly composed of one of the nucleic acids, DNA or RNA, or both. So, too, are genes. The way in which
genes are arranged along the chromosomes of the host cell determines its inherent characteristics and those of its offspring.

Viruses could be regarded as stray genetic chemistry kits which, if
they can purloin a set of extra ingredients from host chromosomes,
can produce many identical new kits (that is, copies of themselves).

In its mature form, a typical virus particle often has a thin shell of
protein which gives it (in electron microscope pictures) the appearance of a quite complex, regular structure. But when such a particle
attaches itself to a living cell and injects its contents through the cell
wall or membrane, it becomes, to all appearances, completely disem-
bodied. The protein shell is abandoned and the strands of nucleic

acid molecules the virus has thrust inside the cell soon become indistinguishable from the cell's normal contents. After a while, however, dozens of new virus particles appear inside the cell, as if from nowhere, and they burst out through the wall or membrane in search of new host cells. The living tissue, of which virus-parasitized cells form part, becomes depleted and diseased.

The process described can be very rapid. Some viruses pass through several generations in a matter of minutes. This destructive multiplication can give rise to a number of diseases, from mild flu to certain types of cancer. But the account I have given, which is modeled on the behavior of one kind of specialized virus — the "phage" particle which invades certain bacterial cells, and which is the favorite subject of most textbook accounts of viruses — tells nothing like the full story.

Death or disease is by no means the inevitable upshot of viral infection. Like other parasites, many viruses have a vested interest in not destroying their hosts; instead, they often remain latent within them for an indefinite period. Such "proviruses" can be transmitted from cell to cell down many generations without once assuming an active or harmful role, unless the host cell begins to function imperfectly for some reason, or ceases to grow and divide. In that event, the virus may resume its active form and burst out of its borrowed accommodation.

In its passive or latent form, a virus may simply knit its chromosome into one of its host's and thus become part of the cell's genetic furniture. Or the virus may replicate independently within the host cell, using a share of the raw materials that go to make up the host chromosomes to cater to its own. Extra copies of the viral chromosome may then be distributed in new cells when the original cell undergoes normal division. Such interference may slightly alter the appearance or biology of host tissues, but it does not directly threaten the host's survival. Biologists are beginning to suspect that this type of latent parasitism may be commonplace in the living world and that much of the DNA in the chromosomes of present-day organisms, the very stuff of their evolution, may have originally arrived in the cell in the form of foreign bodies like viruses which in due course became (to use Allan Campbell's words) "naturalized citizens of the intracellular community." This is a conservative expression of Lewis Thomas's (34) more lyrical assertion that:

We live in a dancing matrix of viruses; they dart, rather like bees, from organism to organism, from plant to insect to mammal to me and back again, and into the sea, tugging along pieces of this genome, strings of genes from that, transplanting grafts of DNA, passing around heredity as though at a great party. They may be a mechanism for keeping new, mutant kinds of DNA in the widest circulation among us. If this is true, the odd virus disease, on which we must focus so much of our attention in medicine, may be looked on as an accident, something dropped.

The fact remains that, when viruses fail to live in harmony with host cells, that "something dropped" can be a very big "something" indeed. Even in their latent form, viruses can alter the behavior of a host in ways which oblige it to harm other organisms. For instance, there is a virus (or something very like a virus) which, when present in some disease-causing bacterial cells, renders them immune to antibiotics. And there are several diseases attributable to viruses in conjunction with simple organisms like bacteria and some fungi. L.H. Rolston and C.E. McCoy(27) described in 1966 a disease of silkworms which was caused by a virus/bacteria combination. Neither the bacteria nor the virus produced disease symptoms when injected into a host in isolation from one another.

The relevance of this discussion of viral parasitism to the question of DED hinges on two questions. Is the virus which is thought to infect *C. ulmi* an ordinary parasite that causes a contagious, weakening disease and so creates the slower-acting waxy form of the fungus wherever it is present? Or are both the aggressive and nonaggressive forms of *C. ulmi* simply different manifestations of the activity of the same viral parasite, or maybe of two competing viral parasites? So little is known of the way viruses behave in general that anybody's guess about the way they work in particular instances is, up to a point, as good as the next man's. For that same reason, it is not altogether fair to speculate too wildly about viral parasitism beyond the ken of present investigations.

In the case of DED, nevertheless, it is reasonable to suggest that viruses, having been recognized as factors possibly relevant to some of the more puzzling questions raised by detailed studies of the disease, are not likely to oblige researchers with clear-cut answers. Less still are they likely to oblige with immediate practical solutions to the problem of the disappearing elm.

We know this much: other fungal diseases of plants undergo mutations similar to those exhibited by DED on account of the interference of viruses. Diseases of cereal crops (particularly wheat rust) had long been known to be liable to change very quickly and dramatically into more and more virulent forms. Agriculturalists protect the future of their crop by planting a race of wheat tested for resistance to attack from the particular brand of wheat rust prevalent in their area. Seed companies also keep hundreds of different stocks of wheat in reserve, each of proven resistance to known, or possible, strains of rust. Such foresight, however, seemed hardly convenient in the case of disease-prone trees. For it can take forty years to establish the resistance to disease of a single tree cultivar, by which time the disease organism may have changed its nature completely and rendered the breeder's innovations useless.*

I have so far left out of the reckoning the most important factor governing the natural history of tree diseases — the inability of trees to adjust rapidly to pressures placed on them by parasites. It is per-

---

*Most recent research into the relationship between *C. Ulmi's* different Jekyll and Hyde forms has considerably altered the emphasis of current thinking, besides painting a much brighter future for elm trees. Work by Brasier and Gibbs (see *Annals of Applied Biology*, 83 (1): 31-37 May 1976), proved the existence of a hybrid form of *C. Ulmi* midway between the aggressive and the nonaggressive strain.

The nonaggressive or waxy form was itself found to exist in two types of slightly differing virulence, but only one of these types, the less virulent, appeared genetically compatible with the aggressive or fluffy type. The appearance and behaviour of the hybrid offspring varied considerably. The confusing results of much of the earlier experimentation could now be understood in terms of the experimental stocks having included the hybrid form. The most important aspect of this improved knowledge was that the hybrid form, no matter what its properties in the laboratory, was much less virulent than the pure aggressive form when injected into elm trees. What was more, it now seemed apparent that the pure aggressive form could not arise from the nonaggressive by a simple mutation. Thus, since the two strains were reproductively isolated from one another, it now seems likely that, where both strains met and mated, the offspring could not be the virulent, fluffy strain but had to take a less aggressive form. The most exciting implication of this research is the likelihood that, in time, the aggressive *C. Ulmi* must yield its place to less aggressive forms of the fungus.

The search for a viral parasite in *C. Ulmi* has been superceded by the new findings, but the reason for the original flare-up of the disease has yet to be established — it is still possible that an explanation may be found in the submicroscopic realm of the virus rather than in the study of DED's geographical distribution around the globe.

Meanwhile the prospect of a truce between elms and DED appears, from the research I have described, to be better than ever before. Unfortunately DED itself has not yet been informed of these developments. The elm populations of Britain and America now stand, at a conservative estimate, at 50 percent of their preepidemic level and are still falling fast. Even if the disease dies out within the next five or ten years the substraction of the elm from the landscape will have gone too far for correction. Meanwhile, fresh epidemics of DED, involving *C. Ulmi's* aggressive form, are hitting other regions — notably Iran and parts of the USSR. If we want to give elms a chance to survive in the truest sense, which means remaining as familiar and useful parts of the natural scene, we must actively preserve elms in large numbers across wide areas and we must do it now.

fectly true to say, of most organisms, that they can generally be trusted to evolve some measure of resistance to parasitic attack, even in the course of a few generations. Thus, though some diseases (bubonic plague among humans, for instance) may appear to threaten certain creatures with annihilation, the threat is very seldom fulfilled. Indeed the odds are that it will be fended off rather quickly as evolutionary time scales go, within tens or hundreds rather than millions of years of adaptive adjustments on the part of threatened organism and threatening parasite alike. The same, however, cannot be safely said of trees. Their evolution is painfully slow at the best of times, so slow, for example, that many trees standing today exhibit features held over from ice age conditions that prevailed over forty thousand years ago and are now virtually redundant. The phenomenon of juvenility[4] is one such feature. Tylosis (see p.79) may be another. Is it a coincidence that temperate zone trees (elm, beech, oak and spruce, for example) which display such features most markedly seem to be more prone to parasitic diseases than trees which show few or none? Or are the trees which have "throwback" features inherently less flexible in their response to disease pressures?

I have suggested that DED may be a "disease despite itself," and not only because of the contribution human commerce has made to its spread. It is also because the elm is, in certain ways, preeminently liable to speed its own destruction. Tylosis, its elaborate natural defense system, kills more effectively than it cures. The habit elms — certainly the most DED-susceptible elms — have of reproducing vegetatively in many regions, and so missing out on the genetic flexibility sexual reproduction can bestow, is yet another cross elms apparently have been obliged to bear. It is yet another brake on their painfully slow progress toward compatibility with their parasitic invaders. Again, the fact that elms abound in highly industrialized countries and are frequently made to grow in the presence of polluted air and soil, or under bright street lighting which upsets the rhythm of their yearly cycle of growth, suggests another reason why they, more than other trees, are the first to succumb. In an already slightly weakened condition, they form a prime target. It may be significant here that factors known to activate innocuous, latent viruses into unpredictable, harmful forms include X-rays, ultraviolet radiation, and certain industrial chemicals (including, of course, certain pesticides). Even quite small fluctuations in normal temperatures (such as will occur in cities) can be a precipitating factor. Any one of these items could conceivably have acted upon the virus which may be present in *C. ulmi*, causing it to change the fungus's most basic nature from within and perhaps specifically sponsoring *C. ulmi's* epidemic strain during the First World War.

Because of the various predisposing factors I have listed, elms may therefore be serving as an early warning of things to come, a warning that no tree is really safe from epidemic diseases like DED.

Now, if the grand evolutionary trends I have speculated upon really are as described, it might seem to follow that responsibility for the problems trees face, at least in temperate parts of the world, cannot fairly be laid on human technology's doorstep. Surely fate, in the guise of evolution, is more to blame than we.

Quite the contrary. Technology has undoubtedly speeded up a drama that would otherwise have taken millennia to unfold. More significantly, technology is in any case perfectly capable of doing something to reverse the present turn of events. In the long term, certainly, there is little that can be done that is not already in hand. The patient, but possibly quixotic, search for a DED-resistant elm cultivar that can replace lost populations goes on and on.

But the day-to-day emergency and wholly necessary efforts to rescue the millions of elms that remain in jeopardy from epidemics of DED remain depressingly parochial. There are ways of minimizing the risk of DED and, where those ways have been put into practice, they have shown remarkably good results. There are even techniques for curing elms already infected with DED. These, too, though not perfect, have given encouraging results. Nowhere, however, has any abiding initiative yet been mounted to preserve elms as a matter of national policy, commanding a due share of national resources.

The real obstacle to any such sorely needed initiative is not essentially a lack of resources but an almost universal tendency to assume that trees are expendable. If one kind of tree (especially if it is not a direct concern of commercial forestry) proves costly to preserve, it is taken for granted that it can either be done without or replaced by another kind of tree, or better still by productive farmland. Not only is this view unsound in terms of biological and ecological realities, it is also demonstrably untrue in economic terms, as we shall see.

Is world deforestation to become, like DED, a disease despite itself, a disease arising not from the actions of viruses or fungi but from misinformation and the clash of vested interests? If not, we must recognize our solemn duty to ask the question: what are trees really worth to us and to our planet?

5 America's Little Hatchet

"At the gates of the forest, the surprised man of the world is forced to leave his city estimates of great and small, wise and foolish. The knapsack of custom falls off his back with the first step he makes into these precincts. Here is sanctity which shames our religions, and reality which discredits our heroes. We have crept out of our crowded houses into the night and morning . . . the incommunicable trees begin to persuade us to live with them and quit our life of solemn trifles. Here no history or church or state is interpolated on the divine sky and the immortal year."

RALPH WALDO EMERSON

". . . I cannot tell a lie: I did it with my little hatchet."
(attributed to) GEORGE WASHINGTON

WHEN NORSE WANDERERS first set foot in America (probably around 1001, on what is now Rhode Island) they stepped unaware into the shadow of the largest undivided forest in the world of that time. In the Americas of those days it may well have been possible to travel from the Arctic almost to the Antarctic Circle without once leaving the shelter of dense woodland, down the length of a continent which — because of its relatively slender breadth — lay open to moisture-bringing ocean winds in almost all its parts; a continent ideal for the growth of forests.

In places our hypothetical journey might have entailed a few detours. There were always Great Plains in the rain shadow of the Rockies, for instance, though there were certainly times when these were nothing like so extensive as they are today. The prairies and arid lands of the Midwest and Southwest were likewise always a fairly steady feature of North America's ancient scenery by reason of similar geographic and climatic pressures. Even the driest of these regions are, however, known to have borne woodland in varying amounts from time to time.

It is likely that, except at a peak period of ice age glaciation, North America was never less than three-quarters forested, during an age which stretched from five hundred to five million years ago. These almost never-ending backwoods were also endlessly varied. Norse pioneers were in a position to witness a goodly fraction of that variety. For then, as now, four completely different major types of woodland lay within a two-week march of Rhode Island.

The vista of dense, rolling woods they sighted on first arrival was presumably part of what is now termed the hemlock-hardwood woodland formation. It now extends discontinuously from southern Canada and eastern Minnesota through the Great Lakes region to Pennsylvania and New England, then south through the Appalachian foothills to Tennessee and North Carolina. This woodland characteristically includes (besides hemlock) sugar maple, red oak, black cherry, yellow birch, white ash and beech, and conifers like white pine, larch and cedar. In an unspoiled condition, it is one of the most visually pleasing forms of woodland outside the tropics. To the wandering Norsemen, it must have struck a startling contrast with the moss-clad tundra around the settlements in Greenland whence they had come.

A little farther north, another kind of vegetation — one very familiar to Scandinavians past or present — was to be found. This was the boreal forest of spruces, firs, larches and pines which still runs coast-to-coast from Alaska to Labrador and the Appalachian peaks, and is essentially a continuation of the northern forest (the *taiga*) of Scandinavia and the USSR. This forest remains the largest in present-day North America, providing the bulk of the continent's timber wealth.

South of Rhode Island, the hemlock-hardwood forest gives way to two further distinctive woodland groupings. Inland from the Appalachians there is the deciduous forest which covers more of eastern America than all its other woodlands put together. Regional conditions of soil, climate and so on conspire, however, to create natural subdivisions of this forest into a number of local types named for their predominant species. Notable among these are the oak-hickory woods which crowd the never-glaciated southern and central sections of the forest's overall area, and the scarcer beech-maple association more typical of northern sectors where the terrain was in fact smoothed out by glaciers in recent ice ages.

Other trees found plentifully in the deciduous forest are elm, ash, birch, basswood, tulip tree, poplar and sycamore. Despite considerable pressure from man and from manmade distortions of the natural order, the deciduous forest today retains much the same mix of species (though, of course, with nothing like the same strength of numbers) as it had a thousand years ago. Certain species — the elm is one — were periodically more in evidence for natural reasons. Others have been more or less eliminated in recent times by man-introduced disease epidemics. Of these, the chestnut provides the best-known example. Until chestnut blight worked its ravages during the past hundred years, chestnut used to form some 40 percent of the tall layers of deciduous forest south of the Ohio River, while oak-chestnut woodland was at least as common as the oak-hickory formation.

Coastward from the Appalachians lies the southern evergreen forest, crowding the Atlantic's edge from New Jersey to Florida. It extends inland, where the Appalachians peter out, from Georgia to Texas. Pitch pine dominates the seaboard, sharing river and stream banks with large enclaves of southern white cedar. Longleaf, slash, loblolly and shortleaf pines form this forest's overall bulk, though evergreen oaks compete successfully with these conifers in several areas. Cypress woods take over from cedars as riverbank guardians in the south and southwestern sectors.

Here the inventory of forest forms theoretically accessible in 1001 to the curiosity of Bjorn and Lief Erickson ends. For our own curiosity, however, we continue this tour of North America's trees. If we are to discuss the future of American forests, we must have a clear picture of their present state.

At Florida's southern tip tropical forest and swamp forest begin to appear for the first time. But the thread of the great South American tropical forests is not actually picked up anywhere else in the USA. Instead it knits in the foothills of the Sierra Madre in southern Mexico with the southernmost fringes of Rocky Mountain forest.

The forests of the American West are a very different proposition from those of the East, for reasons best understood in terms of climate and geography.

Moisture-laden air streaming inland from the Pacific during the winter arrives from the south and southwest. As these airstreams leave the ocean, they meet with a tripwire in the form of a series of coastal mountain ranges. They rise and cool: the water-vapor they carry forms clouds. These soon condense, drenching most of the West Coast with heavy winter rains.

Figure 8

Map of the climatic plant formations of North America, showing forest boundaries of about a century ago.

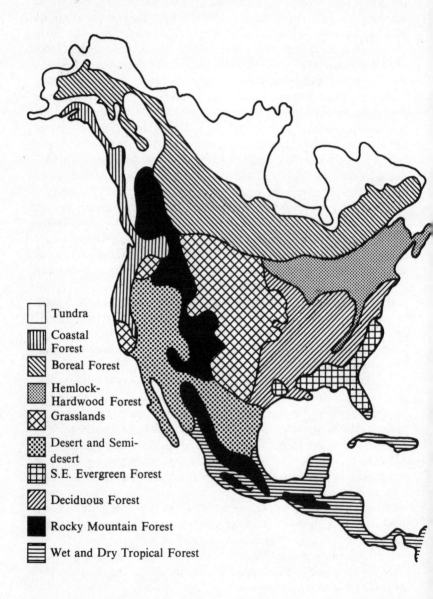

Tundra

Coastal Forest

Boreal Forest

Hemlock-Hardwood Forest

Grasslands

Desert and Semi-desert

S.E. Evergreen Forest

Deciduous Forest

Rocky Mountain Forest

Wet and Dry Tropical Forest

In summer, on the other hand, the ocean winds arrive mainly from the cooler waters west and northwest of America. When they reach the coast, the further cooling they undergo does not suffice to trigger heavy rainfall except to the north, where the coastal ranges are at their highest. Annual rainfall in the latter region can top two hundred and fifty inches. Lush temperate rain forests — seen at their best on the Olympic peninsula in Washington State — therefore clothe the seaward slopes. In this truly olympian treescape, very tall Sitka spruces, western red cedars, Douglas firs, western hemlocks and silver firs tower up together. At the latitude of the Canadian border the same forest also extends inland as far as the Rockies.

Farther south, in northern California, the lower coastal ranges wring only enough moisture from the summer winds to create a narrow belt of cloud and fog along their western flank. Here, dense redwood forests, containing some of the world's tallest trees, loom mysteriously out of the swathe of half-condensed water vapor which will succor them till the next winter's heavy rains begin. Such redwood forests were quite widely distributed across the continent during prehistory and early historic time — a fact which testifies to the profound changes the American climate has undergone in the distant and not-so-distant past.

Near the southern Californian coast, the coastal ranges are too small to temper the weather significantly. Summer droughts are the norm, as the rain-bearing winds cruise by without releasing more than a tiny fraction of their load. Chaparral — a form of drought-resistant, dwarfish scrubland — is the characteristic vegetation here. Even the chaparral, however, like its ecological equivalent in the Mediterranean regions, the maquis, has known better times and richer forms.

To the east of the coastal ranges, in Nevada, Arizona and Utah, occur North America's driest regions. As air masses descend, having crossed those ranges and skipped the intervening sierras, they warm up and begin to absorb moisture rather than release it. Thus the eastern slopes of certain ranges may be desolate semideserts while the western slopes, only a few dozen miles off as the crow flies, burgeon with luxuriant rain forest.

When the dry lands behind the coastal hills and mountains have yielded up their mite of moisture and added it to whatever remains in the wind of the ocean's hoard, the Rocky Mountains rear up to take their share. By this time, however, the potential wealth in the clouds is much less substantial. The Rockies wring from these enough moisture to support forests which are impressive in extent but nothing like so lush as the coast's rain and cloud forests.

Rocky Mountain forest displays several distinct zones where — depending on altitude — different trees predominate, nearly all of them conifers. Arctic tundra on the topmost peaks gives way to Engelmann spruce, which thrives on and above the snowline, often accompanied by alpine fir, alpine larch, whitebark pine and bristle-cone pine. Individual bristlecone pines three thousand or more years old have been found growing in this zone. They are probably the world's oldest living creatures.

Below the spruce zone, Douglas fir abounds, grading at lower altitudes into ponderosa pine woods. The two last-named species are the most common in western America.

East of the Rockies, the Great Plains spread in a virtually treeless expanse, waiting for their meager annual ration of ten to fifteen inches of rain. Their eastern boundary, though much better watered, remains a grassland region mainly by reason of repeated burning. The fires often ignite without human help, but manmade fires have also played a large part in establishing and maintaining the present limits of the plains.

Where the plains end, the eastern forests begin. Their supply of rain comes mainly from airstreams moving up from the Gulf of Mexico. It is thanks to the lack of major geographic obstacles to this airflow that the eastern forests are richer and spread much farther inland than those of the West.

We have now returned to the starting point of this simplified summary of America's major woodlands. Jack McCormick's book *The Living Forest*(16), on which the foregoing account is based, is a congenial source of further information on the subject. What concerns us most of all here, however, is the story behind the way America's superb forests have been reduced in the course of the past few hundred years from about three-quarters to a bare one-third of her land area. Even their much reduced present scale and variety is today insidiously set further at risk.

How, first of all, did the spectacular reduction come about? Answers to this question must be sought in a sequence of history which begins long after the Norse landings of a thousand years ago.

First, neither the Norsemen who founded the first European settlements in America, nor the Eskimo and Indian people whose homelands they intruded upon, were committed to the practice of agriculture as an exclusive way of life. Some farming and herding was carried on in the America of those days. But the forests provided, on the whole, a generous enough ready-made livelihood in the form of game and gatherable plants, along with construction materials, fuel and textiles. It is possible that the Scandinavian newcomers were slightly more interested in tilling the soil than the Eskimos and Indians of the region but, for reasons history does not clearly record, the Norse settlements were in any case short-lived. Certainly they made no abiding impact on the forest. Trees continued to outnumber men by an incalculable ratio. The burning of forest in order to flush game out of hiding was the only activity that then posed any lasting visible threat to the land's giant tarpaulin of treetops.

It was not until five centuries later — about five centuries before the present — that the exodus from the Old World to the New began in earnest. It was this population influx that was to make a truly astonishing impact on the North American treescape.

Spurred on by references in the newly popularized works of classical authors (especially Pliny) to a lost continent in the western ocean, adventurers from southern Europe began to set their course westward in the late fourteenth and early fifteenth centuries. The accidental discovery of the Canary Isles by Genoese traders in the 1390s lent credence to the old tales. Within a half-century Madeira and the Azores had come within the ken of Spanish and Portuguese navigators.

When Zargo discovered Madeira in 1419 or thereabouts, he found it uninhabited and so densely wooded — *madeira* means "wood" in Portuguese — as to impede settlement. The group of colonists who returned about a year later decided to burn down some of the forest on the island's southern slopes. Their fires got out of hand and, according to contemporary sources, burned for more than seven years. So fiercely did they rage during the early months that the would-be settlers were forced on many occasions to take refuge in the sea. This bizarre beacon with its titanic column of smoke far out in mid-ocean could serve as a fitting symbol of Europe's latent intentions toward the New World she was then halfway toward discovering.

Not, however, that all fifteenth-century Europeans were entirely ignorant of the harmful consequences of deforestation. Columbus himself was keenly aware of these perils. He visited Madeira on at least two occasions, once before the forest fires had done their worst and again after the blaze had died. At this point the cloudbelt and the natural vegetation had been forced back to the highest altitudes, leaving the agricultural zone near the southern coast barren through drought and erosion following only a few years of intensive cultivation. When Columbus reached the island of Jamaica, he remarked (according to Hans Sloane who, two centuries later, retraced the great navigator's course by reference to the *Santa Maria*'s logbook) that it: "...surpassed any he had yet seen for ... fertility, victuals, &c., which he judged to come from its being watered by showers drawn thither by the woods, which he had observed to produce the like in the Canaries and Madeira before their being cleared of trees."

No such insight was to stay the hand of the English, French, Spanish and Dutch colonists who settled what is now the USA during the next three centuries.

Originally, much of the space in the holds of the settlers' boats was occupied by axheads. Soon they had the means to make their own iron tools — especially the trade ax, an unsophisticated item by European standards, but one which was much in demand among the American Indians. To the massive land-clearing thrust of the invaders was added a significant increase in the efficiency with which indigenous Americans were now able to exploit their natural surroundings — acting out the apparently universal human urge to exceed the necessary. Here is a poignant analogy with the strategy of the task forces busy at present in Amazonia, spearheading the twentieth-century technological invasion of South American Indian homelands and reserves. Persuasion, much less force, was unnecessary. The uncannily efficient tools and domestic utensils dispensed so liberally by the strange newcomers soon created a form of dependence far subtler, yet often far stronger, than any imposed by hordes of missionaries or flotillas of gunboats.

It is ironic that the first axheads unloaded by the Founding Fathers at the edge of the virgin forests of America had been formed of metal purified over charcoal culled from Europe's exhausted woodlands. What was left of those lands had already been cut over several dozen times in most places. Politicians and rulers in Europe were slowly becoming aware that a scarcity of forest could be a serious handicap to a state's economy and military power. In those salad days of capitalism, deforestation may well have been the greatest single unsolved problem of statecraft, though it was still not fully recognized as such. The present-day tussle for ownership of the world's oil supply may not be too exaggerated an analogy. So the fabled gold and silver hoards of the Aztecs and Incas were, in fact, paltry prizes by comparison with the soil and timber wealth of North

America — wealth easily accessible to any able-bodied settler. While the English wrestled with the French, Spanish and Dutch for sovereignty over the easily farmed northern part of the new continent, another less obvious but equally historic struggle gathered pace at the limits of settlement as wave upon wave of colonists encroached on the forest's measureless shoreline.

Eccentrics and daredevils set their sights far beyond the beaten track, creating, whenever their enterprise succeeded, isolated pools of settlement. These were soon linked to the mainstream of migration by first a trickle, then a flood of imitators. Soon the eastern half of America was crisscrossed with manmade trails. Resources could be moved to and fro with ever increasing ease. By the time the British had won their battle for the freehold of North America, the young colony's economy was booming. But in the Britain of the early 1700s, the economy, like the forests, was in tatters. The iron-producing and iron-using industries were in decline for want of wood charcoal to power the smelting process. Thousands of tons of iron ore had to be shipped daily up the coast and through the Scottish lochs to refineries improvised near the deciduous forests, which were Britain's last major repositories of suitable fuel(5). Coal had not yet been widely discovered and utilized as an industrial and domestic fuel. Trees were still the only important source of energy.

John Evelyn, in his famous *Sylva*, one of the most influential books of its time, awoke the British public to the connection between good forestry and general prosperity. He appealed, above all, to his readers' patriotism by pointing out that the decline in Britain's forests threatened to lead to a decline in her military power and her dominant hold on world trade. Ships were, of course, a country's most important weapon and all were timber-built in those days. But only one tree in dozens was suited for use in constructing the essential main components of new ships. Evelyn's strictures were soon taken to heart by the governing classes. Landowners scurried to restock their straggling woodlands with oaks, pines and many other useful trees.

But the legacy of centuries of neglect was already only too apparent. Only by limiting the right of American colonists to trade in their own vessels with states other than the mother country was a debt-ridden Britain able to compete in world markets with her own booming New World settlements and their brand-new merchant armadas. This was one of the many unfair restrictions which were eventually to lead to the War of Independence.

The nineteenth century saw something of a relaxation of the pressures which had for so long been placed on Europe's forests. The general use of coal for fuel, and iron for ships, gradually became feasible, while the tree-planting fever of the previous century bore fruit in a partially restored countryside. Rotation of crops and other improved agricultural practices saw to it that farmland could now be recycled, obviating some of the need to encroach on woodland in search of further caches of fertile soil.

In America, however, those pressures did not relax for an instant. In the years between 1800 and 1870 the population of the USA exploded from five and a half to thirty-eight million. The consumption of wood followed suit. Wood was used for domestic heating and cooking, for most agricultural and industrial processes, for construction, and as fuel for steamships and locomotives, to the considerable exclusion of any other source of energy or material. Why seek other resources, when the forest was plainly inexhaustible?

At the same time, the clearing of forest for agricultural settlement or to enlarge grazing areas intensified the profligate waste. For every single tree harvested for a specific, useful purpose, thousands were merely hacked down and burned in the general interest of creating and expanding areas of clear ground.

In the more densely populated areas (New England, for example) every available and suitable square foot of land was soon under cultivation. Ignorance, sometimes willful, of crop rotation and soil fertilization techniques inevitably caused the land's productivity to dwindle hopelessly after a varying period of three to thirty years. During the decades that followed the Civil War, many thousands of small farms in New England and elsewhere were abandoned for new, fertile lands in the West, far away across the plains. A desolate waste of overgrown fields and ramshackle homesteads remained in the wake of the great push westward. Only a few homesteaders stayed on to scrape a living from the least impoverished of their fields.

Gradually the eastern forests did reclaim their former haunts, though they never regained their full prime. By the time agriculture was fit for revival in the East, fields and farm units were larger. But the borderline between farm and forest was nevertheless held in much greater respect. The need to sustain forests as a matter of active policy had at last become plain. Demand for timber had, however, never ceased to expand. Yet it was no longer possible, at least in the more developed states, to meet this demand by random logging. Distinct forest-using and forest-producing enterprises therefore took shape rapidly during the late nineteenth century.

These were, from the word go, large scale but labor intensive by comparison with agricultural concerns. Early forest industries in the USA were committed to cutting down ever larger numbers of trees just for the sake of it, or rather just so as to maintain turnover and keep a big labor force fully occupied. But they did also recognize the need to husband forest resources and promote replanting programs, if only to ensure that logging operations did not get stranded too far from the rivers, which transported the harvest to market. The mechanization of sawmills eventually began the series of changes that were to transform big forestry from a labor-intensive into a capital-intensive affair. Yet none of these changes came near to addressing what has always been the most stubborn malaise of commercial forestry — a perverse urge to subordinate biological growth to capital growth.

At much the same time as the excesses of big forestry became obvious in most parts of America, the USA was pioneering the world's first national parks (beginning with Yellowstone in 1872) and was ready to take still more revolutionary steps toward conserving natural, and especially forest, resources. In 1891 Congress approved the establishment of enormous forest reserves, later to become known as national forests. The basic idea of reserves was not new, but reserves on the American scale were hitherto unknown.

The American government had always been a major land and forest owner. Throughout the nation's history, however, publicly owned land was regarded simply as property held in trust until it could be transferred to private ownership in accordance with the free-enterprise ethic. By 1890, the lion's share of the nation's most productive woodland was in the hands of railroad companies and other industrial corporations, a few semiprivate state bodies and, increasingly, gaggles of land speculators. The vagueness of property transfer laws in certain states left scope for racketeering on a grand scale. It was largely in order to prevent such irregularities that forest-reserve legislation was enacted. Yet a dawning awareness of conservation issues also played some part in Congress's unprecedented espousement of the principle of state ownership. The threat of drought and flooding downstream from deforested highlands, for example, was by then well recognized.

Much of the newly reserved forest land was in the West, where many old conifer forests still remained in almost virgin condition, mainly because their position on mountainous terrains had kept them exempt from systematic exploitation. Many more forests were added to the public inventory during the first few decades of the twentieth century, particularly around the time of the Depression, when many a bankrupt landowner had to forfeit large acreages of forest to pay off tax debts.

In 1928 the McSweeney-McNary Act led to the establishment of the Forest Service — a branch of the Department of Agriculture responsible for setting standards of management in national forests, and for maintaining a watching brief on the general condition of America's forest resources.

Meanwhile, the reversion to woodland on the part of run-down farmlands in the East continued apace. Right up until the mid-1950s, this trend upheld the impression that, in terms of overall area, the forests of the USA were prospering. They seemed, moreover, to be doing so against the general trend in the developed world, where a slow but steady loss of woodland had been chronicled ever since surveys had begun.

In fact, most of the bonus provided by the abandonment of run-down farmland was, and is, deciduous woodland with little forestry potential. Most of it is still owned by small farmers who will, if the lesson learned in Europe is anything to go by, part with it willingly as soon as a new use for the space it occupies can be devised.

Against the trend (itself highly reversible) toward the spread of hardwood forests in the USA during the past century must be set the rapid increase in the commercial exploitation of softwoods — by far the most profitable source of construction timber and pulp — during the same period. Such exploitation has escalated so wildly that the balanced rotation of this sought-after crop has been thrown seriously out of kilter. In 1970, for example, forty-eight billion board feet of dead softwood tree were harvested, while only forty billion board feet of live tree were grown — a discrepancy which increases logarithmically year by year. (The term "board feet" refers to actual or potential amounts of sawn timber end product obtainable from a harvest or stand of trees, reckoned on the basis of their size and age. A board foot is actually a piece of wood one foot long, one foot wide, and one inch thick.)

More has been done in America during the past hundred years publicly to encourage forest conservation than in almost any other developed country. The USA and Canada together still harbor a clear fifth of the entire world's timber resources. Yet nearly all of North America's forests still live under direct threat of unbalanced exploitation or straightforward destruction. This contradiction takes some explaining. In the national surveys published by the Forest Service in collaboration with forest industries, we can gather — both along and between the lines — some facts which may help to clarify matters.

An estimated 754 million acres of the USA were classed as forest land in 1970 — about one-third of the nation's total land area. A third again of this subtotal is — in Forest Service phraseology — unproductive timberland. In the case of some Alaskan forests, unproductive simply means inaccessible to exploitation, but most of the land in this category is semiwoodland or scrubby forest in dry regions, enclosing a large unknown amount (certainly up to half the total area) of open space. Even the forests described as productive in the surveys are liable to display a considerable scattering of clearings of one sort or another. Figures denoting the overall area and tree complement of these forests are only statisticians' projections of censuses taken in relatively small and well-stocked sample areas close to access roads. Hence, if our terms are interpreted strictly as properly referring to dense woodlands where grass is scarce or absent), it is doubtful whether the USA is more than one-quarter forested.

Of productive forests, fewer than one percent of the national total enjoys the status of "productive reserved timberland," where timber operations are forbidden by law. A tenth part of America's reserved forests has only deferred status; that is, its protection as a wilderness area, recreation area or whatever is merely provisional. So (somewhat as in Brazil) less than *one percent* of all true forests found in the USA is completely secure from logging or clearing. Unlike Brazil's mainly virgin forests, however, those of the USA (Alaska excepted) have all actually been cut over at one time or another.

The largest category in the Forest Service's analysis of US forests is that of "commercial timberland" — meaning any woodland suitable or available for the harvesting of industrially useful timber. Two-thirds (some five hundred million acres) of America's forest land falls into this category.

Three-fourths of the nation's commercial timberland is found in the East, where hardwood forests (particularly oak-hickory forests) predominate. The more commercially viable softwood forests are divided roughly evenly between West and East. Their combined area matches that of the hardwood forests. It is in the Douglas fir and ponderosa pine forests of the West that the most intensive softwood forestry operations are concentrated.

Almost 20 percent of the commercial timberland of the USA lies within the boundaries of national forests. Forest industries themselves own only about a tenth of the remainder, though they also lease some timberland from private owners, whose holdings form by far the largest sector (about 60 percent) of the national total.

Industry buys timber from the national forests in large quantities. Although the size of America's publicly owned forest reserves has remained fairly constant over the past few decades — about 186 million acres — the proportion of this area which is operated in the interest of commercial forestry has increased swiftly during the same period — from 73 million acres in 1945 to 92 million acres in 1970. Estimates of the amount of timber harvested within national forests have also soared, from 5.6 billion board feet in 1950 to almost 15 billion board feet in 1970 — a rise of over 250 percent.

The stark fact is that, in recent years, the Forest Service and the Bureau of Land Management (another of America's federal forest-management bodies) have come under increasing pressure from industrial interests to allow more wood to be removed each year from the national forests and other federally owned woodlands. Expensive publicity and lobbying campaigns have sprung this demand against the grain of the Forest Service's avowed aim of a multiple use for forest lands. "Multiple use" should mean the reconciling of commercial logging operations (balanced against planned regrowth on a dynamic or "sustained yield" principle) with environmental meeds (care of watersheds, wildlife and so on) and the demand for recreation areas. The forest industries claim that more logging on public land will be unavoidable if projected demands for timber are to be met in future and America's current housing shortage tackled.

Gordon Robinson(26), who for twenty-seven years was chief forester for Southern Pacific's California forests, believes that: ". . . there is growing evidence that the Forest Service has been yielding to pressure by greatly increasing the sale of timber, and is now grossly mismanaging the national forests."

Much of the national forest land most affected by the expansion and intensification of commercial forestry is by no means outstandingly fertile. By national standards there are many mature stands of softwood trees, but these, although profitable, have been allowed — and have needed — a far longer time to grow to a good size than the trees in fertile and well-watered industry-owned areas which are more naturally suited to tree farming. Official estimates of the recovery rate of less-favored national forests after harvest do not take proper account of this and still other important variables.

Robinson suspects that industry's petition for a greater share of national forests is a sign that America's "long-predicted wood famine has finally arrived." Nevertheless, the amount of timber cut and used annually, nationwide, has not increased substantially since 1900. Nor is there convincing reason to expect any future boom in timber demand, provided prices are not kept artificially low. In Robinson's view, if forest industries had taken better care of their own estate, they would not now be clamoring for a bigger slice of the publicly owned softwood forests. For they cannot plead increased demand.

Before sharing Robinson's analysis of the ways in which the Forest Service and forest industries "count oranges and sell apples," juggling published figures and management boundaries, apparently to distract attention from the true extent of commercial operations in national forests, we should first examine the general state of forest products' commerce in the USA from the industrialist's viewpoint. Is America really underexploiting her forests?

The USA harvests, consumes and exports more wood (in one form or another) than any other country in the world. It is also one of the world's biggest wood-product importers, having ceased to be self-sufficient in such products about half a century ago.

Seventy percent of the domestic harvest takes the form of saw logs, destined mainly for use in construction work. An average wooden house uses about ten thousand board feet of lumber. Most marketable sawtimber trees yield five hundred to a thousand board feet of lumber. Even buildings constructed mainly from concrete require considerable additional quantities of timber or plywood for formwork, fittings and so on.

Pulpwood accounts for a further 15 percent of the harvest. This is marketed in the form of logs, chips or pulp in varying stages of refinement, depending mainly on its intended market state as a bulk export material, or as a raw material for home consumption in the production of paper, card, fiberboard, plastics, silvichemicals and so on.

Rather surprisingly, at least 10 percent of America's annual timber harvest still goes up in smoke. Fuelwood is still widely used for domestic heating and cooking — more widely, indeed, than surveys can possibly measure. Miscellaneous products — fenceposts, telegraph poles and the like — form the remaining 5 percent of the total.

In any survey of America's trade in forest products, then, sawtimber and pulp claim prior attention as the most representative and most economically significant.

The US harvest of sawtimber in 1970 was about fifty-nine billion board feet. Almost 80 percent of this was obtained from softwood forests, mainly from those of the West. In the same year, domestic consumption was sixty billion board feet and rising fast. Forest Service projections, assuming a 50 percent human population increase in the USA by the year 2000, forecast a rise in such consumption to about ninety-seven billion board feet by the turn of the next century, if 1970 price levels stay fixed. Even allowing for rising prices, the increase is expected to be enormous — up to seventy-four billion board feet. On the basis of these projections, forest industry pundits claim that, unless there is to be a timber famine and a consequent acute housing shortage in the USA in the near future, the domestic production and the importation of softwood sawtimber will have to be stepped up dramatically.

America's pulp market appears to be in a similar predicament. In 1972 the domestic consumption of paper and paperboard was 64.3 million tons. Forest Service projections imply that this figure will rise by the year 2000 to between 130 and 190 million tons, depending on the way prices behave.

It is on predictions like these that the wood products' industries of the USA base their bid for greater access to the nation's forest.

But how realistic are these forecasts? Is the specter of a housing supply crisis to be taken seriously in view of the fact that the USA *exports* enough sawtimber every year to build almost a million homes? And is the present level of domestic consumption of timber and pulp, the level on which all forecasts are based, true to necessity or rather a result of wasteful and profligate national habits in the same way that the large American car uses more gasoline than is necessary?

In 1959 each US citizen consumed about twice as much sawtimber as the average Russian, four times as much as the average Englishman, and six times as much as the average Frenchman. Technological changes (use of concrete in buildings and plastics in packaging, for instance) have slightly reduced *per capita* consumption of wood in the USA over the past few decades. But it is still easily the highest in the world. Of paper and paperboard alone, in 1976 every single American consumed about four times the weight of an average adult male (see Table 3).

Table 3

1976 per capita consumption of paper and paperboard

	kg/yr		kg/yr
1. United States	267.1	11. Australia	131
2. Sweden	202	12. Belgium	127
3. Canada	199	13. United Kingdom	122.6
4. Denmark	160	14. New Zealand	111
5. Netherlands	142.7	15. France	105
6. Finland	136	16. Austria	94.6
7. West Germany	133.3	17. East Germany	80.5
8. Switzerland	133	18. Italy	76.4
9. Japan	132.5	19. Ireland	76
10. Norway	132	20. Iceland	76

To say that there is a Parkinsonian connection (where demand rises to meet supply) between the amount of forested land a nation boasts and the average amount of timber its citizens consume, seems to be stating the obvious. Pulp products, however, seem particularly out of place in such an equation. Russians, for example, are big consumers of construction timber. But Russia does not appear on the list of the top twenty countries in terms of *per capita* consumption of paper during 1976, despite the fact that Russia was the world's fourth-largest paper *producer* in that year.

Certain countries, notably Japan, ship in large quantities of pulp or pulpwood chips from the world's richly forested countries and process it themselves on a very large scale. Japan and the UK have in fact been buying almost all of America's pulp, pulpwood and sawtimber exports during recent years. So the two countries in Europe and Asia most similar to the USA in culture and marketing conventions display the same disproportionate greed for imported forest products. Coincidence?

High-pressure marketing has pushed consumption of wood — particularly of pulp products such as packaging materials and bookpaper — close to saturation point in the USA. By comparison with European producers, the US pulp industry reclaims less of its colossal output in the form of reprocessed wastepaper than it might be expected to if it currently really faces a raw-material famine. In 1976, the wastepaper recovery rate (consumption of wastepaper divided by consumption of all paper and paperboard) was over 32 percent in the EEC and around 25 percent in the Comecon countries. The USA lagged behind with a rate of under 22 percent in the same year.

The average American's enormous appetite for wood products and timber is, to a great extent, a straightforward result of the spectacular efficiency of production and marketing operations in the USA. By consuming wood products of lower quality, in smaller quantities and at higher prices, Americans could in fact easily restore their country to self-sufficiency in wood. But such sacrifices are, of course, the exact opposite of the target US producers — quite legitimately, in their way — aim to achieve.

Now, it should not be forgotten that high levels of production of — and vigorous stimulation of demand for — forest products has been achieved in the USA and elsewhere by concentrating commercial operations on softwood raw materials. Trees from big softwood forests are much easier and cheaper to harvest, process and regrow than trees from scattered hardwood forests — such as represent more than half of America's current potential timber supply. Consequently, while cutting continues regularly to exceed growth in America's softwood forests and plantations, hardwood trees rise relatively faster than they fall, in comparison to softwoods. This does not mean, however, that the hardwood forests are still increasing in area. Since the late 1950s their absolute domain has been shrinking quite rapidly. Currently it diminishes by perhaps as much as five million acres a year.

Meanwhile, in the softwood forests, despite the biological thumbscrews applied by producers in the form of increasingly large harvests and increasingly brief rotation periods, productivity has actually greatly improved during the past thirty years. This is thanks to better fire and disease control and to more intelligent use of logging residues but, alas, not on account of any net increase in the bulk of forest planted.

In 1959 fungal and viral disease, insect infestation, fire and storm damage caused the death of commercially viable trees to the tune of 131 million board feet and stunted the growth of surviving trees by 406 million board feet in North Carolina alone — thereby depriving industry of a total of 537 million board feet of wood. Careful attention to such problems gradually improved net growth of commercial timberlands nationwide — by about one-third between 1950 and 1970. The byproducts of logging and milling, which can amount to a quarter part or more of the timber cut, have also been used much more efficiently during recent years, creating market gains of similar dimensions.

Against these market improvements must be set the continual reduction of rotation cycles and soil fertility brought about by the intensification of commercial timber exploitation. Old-fashioned forestry used to leave logging residues to decompose where they lay. Though no doubt wasteful in commercial terms, this practice restored nutrients to the soil in substantial measure. Only the least potentially soil-enriching part of the tree — its trunk — was withdrawn. Again, the disturbance of forest soils by modern logging and reseeding machinery encourages further nutrient leaks. There is no getting around the fact that intensified big forestry endangers the security of America's conifer woodlands — especially in the West — from disruption and visible denaturing.

It seems reasonable, on the face of it, to ask why the vast broadleaf woodlands of the American East are not drafted into use to take the pressure off the nation's softwood forests. One trouble is that the production procedures normally used in commercial forestry — the mass harvesting and planting techniques, the battleship-sized processing machines — have become so specialized in coping with softwood materials that the switch to hardwood production without killing demand on home and foreign markets would require an entirely new technology.

A further major problem is the pattern of ownership of hardwood forests. Partly because of having been neglected for some time as commercial propositions, many of these forests are owned, not by a few large organizations, but by a multitude of small farmers and landowners with their own (perfectly sensible) ideas about the way they wish their property to be used. Financial incentives to farm timber along the lines forest industries recommend often do not outweigh the drawbacks to investing in so cantankerous a crop. Nor do they outweigh the advantages of clearing woodland altogether to create new farmland or housing developments, or of leaving them intact and promoting them as recreation areas. There still remain quite large areas of softwood forest in the East which, like the hardwood forests, never come to market on account of such ownership patterns and preferences. Natural succession, however, sees to it that unmanaged pure softwood tree stands in the East are slowly but surely transformed into commercially unprofitable oak-pine mixtures. Without active management techniques applied to them to pre-

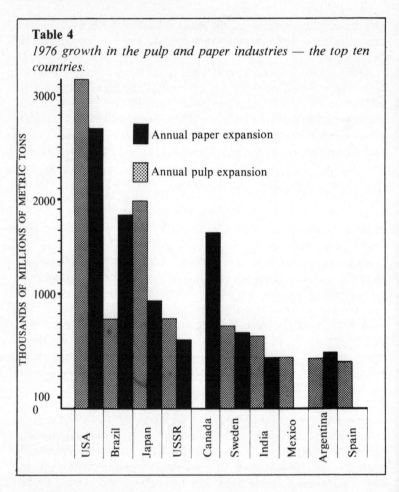

Table 4
1976 growth in the pulp and paper industries — the top ten countries.

vent their own dense shade and acid soil from discouraging seedling growth, they yield their place to more versatile trees.

Unwilling or unable to develop a domestic supply of hardwood timber, American industry has looked more and more to other countries to correct its softwood deficit. Canada bears the brunt of this drive. In 1970, 95 percent of US pulp and timber imports came from Canada. This percentage represents well over half Canada's total world wood exports, half her total production of lumber and 60 percent of her total production of pulp and pulp products.

Curiously, Canada is also the main competitor with the USA for world export markets in wood and wood products. Recent declines in the value of the US dollar have enhanced the growth of US exports in this field while inhibiting that of Canada. Yet, without the help of imported Canadian pulp and timber, the USA would scarcely be able to keep home markets supplied, let alone compete on world markets. As Table 4 shows, pulp is a high-growth industry in Canada. Yet home production of paper and other finished pulp products shows little or no growth in 1976. Geared as it is to the needs of US consumers and the value of US currency, the Canadian paper industry has to mark time during periods of economic hardship in the USA, waiting until the Canadian dollar catches up with the downward trend of the US dollar's value. Meanwhile, pulp producers and processers in the USA enjoy a stint of export expansion.

From the same table it is also noticeable that Brazil is increasing production of both pulp and paper at a rate second only to that of the USA. Behind this statistic lies the fact that tropical South America is fast becoming, like Canada, a reservoir of forest products for the world market. Unlike Canada's, however, Brazil's harvest takes the form of mainly hardwood timber and hardwood-derived pulp. Low labor costs, very rapid biological growth rates, very high yield per acre, and shamefully slack legislation in respect of ownership and management lend Brazil's broadleaf forests a profitability to which the hardwood forests of the USA could not possibly aspire. Though slow hitherto to take advantage of wood supplies from tropical America, the USA will amost inevitably soon increase its consumption of construction timber from this source. And, no doubt, consumption of pulp, too.

The world is at least partly aware by now of the environmental problems which big forestry on the US model can cause in the tropics. Less well known is the impact it has on the boreal forest of Canada. Several ecologists (among them G.F. Weetman) have voiced a fear that commercial demands made on Canada's forests (few of which remain in virgin condition) have seriously jeopardized the long-term stability and productivity of these apparently very robust ecosystems.

When we consider the way the USA nowadays reaches with both hands beyond its own shores and borders for sawmill and pulpmill fodder, do we not see a resemblance to eighteenth-century Britain's frantic grab for the wood resources of Scotland and the American colonies, once her own woodlands had been mismanaged to a standstill? The USA is certainly making a fair imitation of that grab; Canada is her Scotland and Amazonia her "newfoundland."

Individual Americans will have to decide for themselves whether or not it is possible or desirable to slow down or reverse the runaway growth of timber and pulp consumption in their country. The volume of trade is, after all, up to the individual consumer in the long run.

From the point of view of production and conservation ethics, anyone can see at once that a number of practical changes could and should be made.

For a start, the federal forest-owning bodies should be firmly held to their promise to manage the public forests along sustained-yield, multiple-use lines. These concepts must not be allowed to become mere window dressing or lip service. All forest suitable for setting aside as wilderness should be so classified, unequivocally and permanently. Though a late starter in the field, the USSR has recently been doing just that at an impressive rate and on a grand scale. *Zapovedniki* — natural biosphere reserves equivalent to America's wilderness areas and quite distinct from Russia's three vast national parks — now cover over seventeen million acres of the Soviet Union. Though formerly prey to sudden changes of status resulting from shifts in official attitudes toward their function, the *zapovedniki* now appear to have a stable future and are being thoughtfully developed for nonmarket purposes, while their forest fraction is protected out of down-to-earth caution — with a view mainly to securing the richest possible variety of indigenous trees and other wildlife in the very long term.

In this instance, the USA might do well to follow the USSR's lead, for the reasons behind the creation of the *zapovednik* system are almost certainly based on economic as well as on ecological foresight. It is by no means out of the question that a time may arrive in the future when, as in the seventeenth and eighteenth centuries, a country's power and influence could be measured in terms of its forest holdings of all kinds.

In the private sector — the mainstay of the timber market in the USA — there is also a need for government initiatives to promote good forestry, no less than on publicly owned land. This theme is taken up by Gordon Robinson, who asks:

> ... what is good forestry? It is what foresters had in mind when they offered "multiple use" as a slogan to describe the policies of the US Forest Service — policies to protect the watershed, wildlife, range and recreation.

It takes timber to grow timber. It is not enough to have orderly fields of young trees varying in age from patch to patch. That is tree farming, not forestry.

In looking at a well-managed forest, one will observe that it is fully stocked with trees of all sizes and ages. It will be obvious that the land is growing about all the timber it can and that most of the growth is valuable older trees. It will be evident that no erosion is taking place. Roads will be stable and attractive; they will look laid on the land rather than cut into it. The soil will be intact; the forest floor will be covered with leaf litter and other vegetative matter in various stages of growth or decomposition. This absorbent layer holds rain and melting snow while it soaks down into the ground through animal burrows, pores such as worm holes, channels dug by ants, and tracks left by decaying roots of past generations of vegetation. The forest becomes a vast reservoir of water that gradually seeps down through the earth and comes out in springs — clear, cool water. This is how the forest stabilizes stream flow, and this is what is referred to when one reads of the forest protecting watersheds.

One also observes in the well-managed forest that there are frequent small openings stocked with herbs and browse — food close to shelter for wildlife. Finally, one observes that such a forest stays beautiful and will continue to serve our recreational needs so long as it is so managed.

Good forestry means limiting the cutting of timber to the amount that can be removed annually in perpetuity. Good forestry involves cutting selectively where it is consistent with the biological requirements of the species involved. In other cases, good forestry involves keeping the cuts no larger than necessary to meet the biological requirements of the species. Good forestry keeps the full range of naturally occurring species of plants and animals. And — very important — good forestry allows a generous proportion of trees to reach maturity before being cut.

In his article "Forestry As If Trees Mattered"(26), from which the above passage is quoted, Gordon Robinson exposes some of the sleight-of-hand used by commercial timber concerns to flout the actual law and other less formal guidelines, while apparently playing the game and sticking to the rules. In these conjuring tricks government agencies can sometimes connive, bowing no doubt to relentless pressure from the timber lobby.

For example, in 1945 the Forest Service published its first official survey. This showed the existence of 73 million acres of nationally-owned commercial forest. The second survey in 1953 showed 81 million acres of such forest. The third in 1963 showed 91.5 million acres.

In this time, however, there had been no change whatever in the total area of all forest. The additions had been achieved by reclassification. Land that had hitherto been classified, protected and managed as nature reserve was now suddenly commercial.

A further ploy has involved steadily increasing the size of the working circle. The "working circle" defines an area in a national forest which, having been cropped, may not be cropped again for a specified period. By making the working circle simply into a statistic, and considering the total forest as one working circle, the following becomes possible.

Suppose a given part of the total forest is inaccessible to machinery and hence unworkable. Nevertheless, the permitted tonnage or cubic footage that may be harvested from that whole forest includes the inaccessible area. The harvester therefore takes more (he either literally takes more, or takes it more often) from the accessible areas. But the accounting returns for the forest, which report total figures for the whole "working circle," apparently show that the rules are being observed.

Changing the definition of the permitted harvest, a ruse employed more than once, secretly allows greater exploitation. To begin speaking, for example, of "cubic feet" of timber, as opposed to "usable board feet," allows saplings only a few years old to be legitimately harvested. For these can be pulped — to give cubic feet of timber product — although they could not possibly be used for construction purposes.

What is wrong with taking immature saplings? We have already had many of the reasons throughout this book. In summary, by harvesting the immature tree one has prevented it from making anything like its full contribution to the ecology of the soil in which it grows, to the wildlife which depends upon it, toward sustaining the natural composition of the atmosphere, and to the rain-evaporation cycle.

Allied to the harvesting of immature trees is the practice of clear-cutting. In this process, instead of the selective felling (and planned regrowth) of trees across wide areas, the forest cover is simply ripped off *en bloc*, leaving large patches of land quite bare. A total living matrix has been annihilated. Regrowth on clear-cut patches runs into stubborn problems — competition between replanted seedlings

Figure 9
Stages of natural reclamation of clear-cut land (secondary plant succession)

1 *Weeds, grasses and small shrubs begin to colonize the bare land*

2 *The ground cover becomes more dense, and more shrubs, bushes and pine saplings appear*

3 *After about one hundred and fifty years, conifers entirely dominate the area, but the trees cannot reproduce in their own shade*

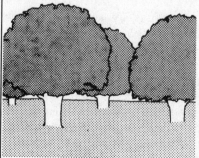

4 *At the climax, up to a thousand years after felling, the conifers disappear and broad-leaf (e.g., oak/hickory) forest becomes the dominant plant mix*

and unwanted "weed" tree species is the commonest of these (see Figure 9); the use of herbicides to solve such problems adds further injury.

The timber concerns are harvesting less (in terms of the size of each tree), but harvesting it faster. For the moment, therefore, production is up. This irrational procedure will work until the over-visited well eventually runs dry.

Robinson finally lists several principles which, he feels, ought to apply beyond question in the planning of any forestry enterprise: (1) rigorous application of the "sustained yield" ideal, so that the tree harvest never removes more of the standing crop than planned regrowth can replace instantly; (2) long rotations, allowing trees to mature to a size where they can yield the best quality timber, rather than immature lots fit only for pulping; (3) "selection" management as distinct from clear-cutting, so that openings in the forest do not become large enough, long enough to pose a threat to the whole forest's characteristic biological qualitites; (4) the maintenance of ecological balance, for instance by leaving some strategically placed trees to mature and die naturally and so provide habitats for wild animals; (5) preservation of locally distributed tree species, rather than their replacement by exotic species not acclimatized to the weather and pest conditions which prevail in the region; (6) protection of the soil. Commercial forestry concerns are well aware of the improved understanding of the dangers of soil depletion which ecologists have achieved in recent years, but tend to regard this lesson as one they need not take to heart while the logs and the dollars keep rolling in. Such shortsightedness is to nobody's advantage in the long run.

From this list of desiderata, one major item is so far missing. What of the hardwood forests of the East, Cinderellas of America's commercial timberlands? Official predictions show a slight increase in hardwood utilization during the next thirty years, as softwoods become harder to come by, but no letup is envisaged in the current obstacles to exploitation of eastern broadleaf woodlands on a commercial scale — too many small forests, or big forests divided up into small ownerships, too many extra production costs and so on. Instead, the replacement of large broadleaf forests by conifer tree farms seems to be the industry's fixed intent.

Yet the days of big forestry may indeed be numbered. The increasing cost of fuel and metals is a trend that cannot be reversed, and it is quite on the cards that small, labor-intensive forest operations, mounted by farmers as a sideline to conventional farming and grazing, will become more and more economically feasible as time goes by. It is also true that myriad small forests have distinct advantages over a few big forests in terms of their usefulness as recreation areas, carbon dioxide sinks, wildlife preserves, and in all respects save perhaps that of watershed protection. A mosaic of forest with other vegetation is, moreover, substantially easier to protect from fire and epidemic disease than a single big forest.

The mixed nature of the hardwood forests of the East is viewed as a major drawback to their commercial viability, but this, too, can be viewed as an enormous potential advantage. The connection between a forest's species diversity and its stability in the face of epidemics cannot be relied on as more than a general tendency. But this is not, as it seems to have been a fashion among ecologists to suggest of late, a useless oversimplification or, worse still, mere wishful thinking. That fashion was begun by R.M. May's contradiction of the traditional "diversity = stability" hypothesis(19). May corrected the mathematical models previously presented in support of this hypothesis to show that, on paper, the larger the number of species in an ecosystem, the less stable that system should be. Though May's view found wide acceptance among ecologists, it was not until very recently that workers like S.J. McNaughton(18) actually applied the reverse hypothesis proposed by May to real ecosystems and found that, broadly speaking, the species which have a share in relatively diverse ecosystems tend to group into "guilds" according to their differing claims on the resources within the system. Within these guilds competition tends to be small — though competition between guilds may be intense. On balance, this pattern of competition places less strain on the stability of the ecosystem McNaughton observed (in fact, a grassland area) than the simpler interactions taking place in comparable areas where species diversity was low.

What is the relevance of these concepts to the future of American forestry? I should like to suggest that the two greatest obstacles in the path of commercial forestry as it is practiced today in the USA are the threat of epidemic disease in big forests and the threat of an oil and metal shortage throughout the world. By the year 2000 these problems will, in my view, make it very difficult for America's forest products' industries to survive, let alone expand, unless dramatic changes are made now in the whole approach to the growing, harvesting and marketing of wood. Western America's splendid natural softwood forests must be conserved for watershed protection and other nonmarket reasons, and for the sensitively managed production of prime sawtimber. But for the production of pulp and other, less vital commodities, America should look to her vast reserve of hardwoods.

Instead of encouraging the abolition or lumping together of small ownerships, the US government should seek ways to stimulate the proliferation of small-scale forestry operations, managed in harmony with farming and ranching, by farmers and ranchers who will bring their wood crop to market in much the same way as any other crop. By easing the tax barriers which now discourage farmers from contemplating wood production, an immediate step toward this end could be taken. Furthermore, instead of seeking to make forest owners scrap the less commercially viable tree species growing on their land, the maintenance of species diversity should be rewarded by added tax benefits. It may, at first glance, seem overcautious or foolhardy thus to reverse conventional trends, for fear of eventualities which will not arise for fifty years, or may never arise. But this degree of foresight is always necessary where trees are concerned. Any major change in our attitude to forestry is a change made for the benefit, or to the disadvantage of, our children and grandchildren. We cannot have things any other way.

Frank Frazer-Darling, in *Wilderness and Plenty*(5), muses that: "The onslaught of the nineteenth century on the forests of North America was so shocking that I have the feeling that that was the reason for the early rise of a sense of conservation in that country." Maybe by now the shock has worn off. Conservation groups are fiercely active on the forest's behalf in some isolated instances, but the greater part of the American public appears to be massively complacent about the fate of their country's still remaining (and still incredible) forest wealth. It is a wealth which cannot be measured in financial terms alone, but which, even in those terms, is the envy of the developed world. Everybody — nature lover, patriot, businessman or whatever — has a personal stake in the shape of things to come where America's forests are concerned. The average American's passionate fondness for his wooded landscape is not in question. American literature bears ample testimony to a transcendent sympathy for forest life and forest atmosphere. What now seems to be missing is a common will to discern amid the jungle of statistics and technical jargon, any sharply focused vision of the way Americans want American forests to look to future generations — the will truly to see the wood for the trees.

6 Fall into Winter

"The saws cutting the huge logs ground out their shrill lament all day long. First you heard the deep underground thud of the felled tree. Every five or ten minutes the ground shuddered like a drum in the dark at the hard impact of crashing rauli, maniu *and larch trees, giant works of nature, seeded there by the wind a thousand years before. Then the saw sectioning the bodies of these giants struck up its whine. The metallic sound of the saw, grating and high-pitched like a violin, following the obscure drum of the earth welcoming its gods, created the tense atmosphere of a legend, a ring of mystery and cosmic terror. The forest was dying. I heard its lamentations with a heavy heart, as if I had come there to listen to the oldest voices anyone had ever heard."*

PABLO NERUDA
Memoirs

IN JUNE 1970 President Medici visited Brazil's densely populated northeast region to see for himself the effects of a drought — the worst for thirty years — which then threatened a wretched end to the livelihoods of millions of peasant herders and farmers.

Very soon after his return, the President made a decision which may one day come to be regarded as one of the most tragically hasty wrong turns of modern history. Within only ten weeks, work had begun on an incredibly ambitious highway building project. Its aim : to open up and exploit the green immensity of Amazonia, a practically uninterrupted expanse of virgin tropical rain forest roughly the size of India and only slightly more populated than the Sahara Desert. The declared purpose of these new highways was threefold. One, to make Amazonia's timber and mineral wealth available to the booming industrialized regions of Brazil. Two, to establish direct land links with and between the neighboring countries of Colombia,

Peru, Bolivia, Guyana and Venezuela. Finally, and ostensibly above all, to create fresh homelands for the half-starved subsistence farmers of the overpopulated northeast.

Birth control was ruled out as a possible alternative solution to the northeast's problems of famine and overcrowding, partly on religious grounds and partly in accordance with the government's view that the country as a whole was underpopulated. Implicit in this last view was the assumption that if Amazonian soils could sustain a million square miles of rain forest (the richest, most massive land vegetation on earth), they could equally comfortably support intensive agriculture and ranching. It was further assumed that if dry-country farmers could eke an ingenious living out of the dusty scrublands of the northeast, they would make light work of raising crops and cattle on land cleared of lush forest.

No steps whatever were taken to test the logic of these assumptions. No ground surveys or environmental studies were considered before the bulldozing began. Such studies were not even made later when it became obvious that the reasoning and planning which steered the project were almost comically faulty. So, for example, long stretches of the new road had to be abandoned because they had been ignorantly routed across areas which lay deep under floodwater for half the year. Each day brought news of similar unforeseen obstacles.

Yet the project continued in the same mad haste as it had begun. In seven years over seven thousand miles of new highway (much of it unsurfaced or ending nowhere) had been added to the map. One road was planned to run (and now runs) directly across the Xingu National Park, homeland of many indigenous Amerindian tribes. Despite public outcry, led by humanitarian interests and anthropologists in Brazil and all over the world, the road ran on. Members of indigenous tribes in Xingu and elsewhere were, and are now being, overtly robbed of the right to pursue their long-established and finely adapted traditional lifestyles, a right which is guaranteed by the Brazilian constitution. How much this reminds us of the North American Indians! The Indian population of Brazilian Amazonia — estimated to have numbered about a million in precolonial times — now stands at under fifty thousand and is continually decreasing. More than half of the 171 official tribal areas set up under Brazil's constitution will be crisscrossed with major arterial roads and European-style settlements by the time the whole highway scheme is finished.

Agriculture and settlement are now encouraged up to a hundred kilometers on either side of all existing new highways. But in the pioneer communities which have already set up fences in Amazonia, the physical side effects of deforestation give plentiful evidence that the highway project's main avowed aim — the creation of a livelihood for immigrant dry-country farmers — was founded on a tragic fallacy.

The immediate environmental impact of deforestation is a cycle of drought, flooding and erosion, with further indirect loss of trees for want of the right conditions to encourage natural forest regrowth. Those catastrophic side effects inevitably obstruct, and are in turn then made worse, by attempts to establish traditional farms, ranches and settlements on the cleared land.

Pockets of soil inherently fertile enough to support settled agriculture do, in fact, lie under a few acres within the tropical forest. But such areas add up to only about fifty thousand square kilometers. The figure may sound large by many standards, but it actually equals only 1 percent of the total area of forest. The rest of Amazonia is largely unsuitable for agriculture, certainly for the kind of agriculture that presently lies within the scope of farmers from Brazil's northeast, even more certainly for mechanized, big-scale monoculture on the North American model.

Rain forest soil is normally deficient in nutrients. The forest has really only itself and the climate to thank for its astounding luxuriance. The protective forest canopy filters the torrential equatorial rain and the harsh sunlight down to the smaller plants in controlled doses. High concentrations of carbon dioxide are conserved in the forest's lower stories, so this undergrowth can photosynthesize efficiently even in dense shade. Most of the nutrients that are cycled within this marvelously complex ecosystem originate not in the soil but in the rain and dust that constantly sift downward through the towering vegetation, to be usefully consumed and transformed at every centimeter of their descent.

Once the canopy layer is removed, the system is irrevocably damaged. Cooked by the sun and battered by torrents of rain, the forest soil quickly loses whatever stores of fertility and resilience it once possessed. It now washes in swathes into the riverbeds, or sets as hard as slate. Coarse weeds now invade it, evicting other less versatile plants. Soil microorganisms are decimated and cease to play any significant role as agents in the recycling of nitrogen, sulfur and other elements indispensable to plant growth.

Wild animals, for their part, are driven away from the vicinity of settlements. Each single departure subtracts a factor from the forest's nutrient equation. For a variety of reasons, domestic animals fail to substitute for wildlife as go-betweens in the process of plant decay. For instance, animal wildlife has a far greater diversity than any domestic farmstead. One result of that diversity is that far more kinds of vegetable food are recycled, as well as the dead or eaten bodies of far more kinds of animals and insects. In addition, pound for pound of body weight, many wild animals recycle more vegetable and other food matter than domestic animals.

With fewer and fewer pest-predators in evidence as habitats become scarcer, pests and harmful parasites thrive. These combine with weeds and soil-degrading processes to cause the failure of settlers' crops, and now add poor nutrition to the many public health problems that arise in newly settled areas. So highway building in rain forest not only prompts deforestation and its attendant ills, but also of itself encourages the spread of human disease epidemics. Warnings were sounded on this score by biologists who as long ago as 1974 pointed out that the planned Perimetral Norte highway would pass right through the center of an area then Brazil's only serious troublespot for onchocerciasis (river blindness). This fly-borne parasitic infection has long been a major public health problem in other tropical areas of the world. Thanks to the distribution routes provided by the new highways, river blindness is now epidemic among ten major Amerindian tribes, and is spreading fast among indigenous peoples and settlers alike. Bilharzia (elephantiasis) and malaria are other epidemic diseases known to have been spread or intensified by highway building and subsequent attempts at traditional settlement. Bilharzia is spread by a freshwater snail, which was kept in check by the acid water of South American rain forests. Once the land was cleared and fertilizers applied, the rivers became alkaline. As a result, the freshwater snail underwent a population explosion in Amazonia.

Yet another unforeseen result of the creation of the highway network has been a series of political disputes between Brazil and some of her neighbors. For Brazilian nationals have taken advantage of the new roads to encroach on more viable land in foreign territory, rather than face the odds of settlement in Brazilian Amazonia. Such unneighborly activities prompted other governments to put a stop to the completion of link roads between Brazil's transport routes and their own. So a few more of the superhighways ran to ground — in the middle of nowhere.

DEATH OF TWO WORLDS?

Though highway building has done much to accelerate deforestation in Brazil, it would be wrong to suppose that the destruction of primary forests in that country is anything new. Brazilian forest trees have been felled at a staggering rate throughout the present century. The state of São Paulo, for example, lost two-thirds of a 60 percent tree cover between 1910 and 1950, while Paraná state lost forest at a rate of 3 percent a year throughout the 1950s. Until very recently, Brazilian government spokesmen hotly denied that unscheduled deforestation was going on in Amazonia itself. But their tune has changed since satellite photographs plainly show that over four hundred thousand acres of forest have been cleared in Amazonia during the last few years, and without any kind of official notification or planning. The global impact of deforestation on this scale in Brazil and elsewhere is likely to be serious indeed, in very many respects. Recent work on the global balance of atmospheric carbon dioxide, for example, provides a reasoned warning that deforestation, together with the burning of waste forest and forest-cropped firewood, now equals fossil fuel burning as a major source of surplus carbon dioxide in the atmosphere. Patterns of climate all over the world — not just in South America — risk severe disruption by the enhancement of the greenhouse effect (see p. 36) that must result if the surplus continues to build up at present rates.

As mentioned earlier, massive deforestation also poses threats to the global availability of oxygen, water and soil.

It was calculated in 1966 that the native vegetation of the United States produces less than enough oxygen to make good that used in the combustion reactions incidental to industry, transport and domestic fuel burning throughout the nation. In other words, the atmosphere over the USA has to be constantly topped up with oxygen produced by marine plankton, and by forests and grasslands outside its national boundaries — if there is still to be enough left for breathing. It is, however, notoriously hard to fill out the proposition that deforestation in the tropics jeopardizes the world's oxygen supply with agreed facts and figures. The issue has been the subject of intense scientific controversy for several decades.

But although no final, detailed line can yet be drawn by the experts, the general public is free to take its own attitude to the controversy. It seems reasonable to say: "I am interested to know for certain how great the risk is or not. But I am not interested in finding out the hard way, by letting deforestation go to an irrevocable point before final measurements can be made, or before my health and my family's health become the measuring instruments."

Man's influence through massive deforestation on the global availability of water and soil resources is of scarcely less concern than his impact on the atmosphere.

Fresh water constantly leaves the land in the form of discharge (mainly rivers), runoff and evaporation. It is restored to the land as rain, snow and other forms of condensed water vapor, themselves the products of evaporation from land and sea. A certain amount of water is retained at a constant level in the ground. It is the renewal of this residue that, more than any other single factor, makes normal agriculture possible.

The evaporation/precipitation cycle, driven by colossal amounts of solar and wind power, draws most of its raw material from the oceans — but most of the end product then falls straight back where it came from. In forested equatorial regions far from the coast, as much water circulates from land to air to land as by the land-river-sea-air-land route and its variants.

Now, man's global impact on the circulation of water should theoretically be small, in view of the fact that nearly all water available on earth eventually finds its way back into atmospheric circulation. No significant amounts are lost into space, and relatively very little is frozen into long disuse at the poles or fixed elsewhere in other ways into chemical compounds. But in practice human activities do in fact have a considerable effect on the rate at which water circulates and on the pattern of rainfall distribution.

Deforestation, especially if it is succeeded by attempts at permanent agricultural or industrial development, is the most important single example of such activities. As noted, deforestation leads to floods and so to soil erosion. The subsequent baking or panning of surfaces leads to increases in the runoff rate of groundwater, and the rate at which topsoil is washed into rivers during normal rain. Agricultural exploitation of deforested land, while it leads to an overall drying out of the soil, nevertheless restrains the normal evaporation of groundwater into the air. Instead, it encourages more moisture to leach off the land into rivers where, especially if the rivers are simultaneously tending to silt up with topsoil dregs, the water takes longer to find its way back into atmospheric circulation.

It is not usually appreciated that deforestation also leads to a general heating up of the land. This effect, especially in equatorial regions, means that less normal rain is wrung from the atmosphere as the result of the difference between air and ground temperatures. Instead, convection rainfall, whereby rain condenses from storm clouds when they come in contact with the cold upper atmosphere, becomes more common. These intense, localized downpours (of course far more likely to cause flooding) thus take the place of the widespread, continual mist of warm rain that is more typical of rain forest weather.

The wrong use of potentially productive land has far worse consequences than its disuse. In particular, the conversion and the use of Amazonia to intensive crop and cattle raising is patently wrong.

There are in fact ways in which Amazonia could become a valuable food-producing area without recourse to mass deforestation. R. Goodland and J. Bookman(6) identify four distinct ecosystems within Amazonia, each suitable for a different type of balanced agriculture.

Varzea is annually flooded land in river basins, where the soil is regularly enriched by silt deposits and so could support a wide range of annual crops, including rice and maize. About 1-2 percent of Amazonia's area can be described as *varzea*.

Then there are the pockets of flood-free but fertile soil (1 percent of the total Amazonian forest area) which could survive conventional agriculture or even limited cattle ranching.

A few pockets of *cerrado* (a type of grassland ecologically equivalent to African savanna) also occur within the forest. They tend to be based on a more stable soil than that found in the forest proper. Here, too, limited ranching and the raising of crops (particularly soy beans) can be quite feasible.

In the mass of vulnerable forest itself:

... some types of agriculture can be sustained, but only if four related precautions are respected. First, the closed nutrient cycle must be maintained. Nutrients must not be allowed to leach into the soil. Second, the forest canopy must not be perforated. Third, the extent of nutrients imported into the ecosystem as rain, solutes, dust, fixation by plants and a little from the substrate, should be used to determine the sustainable size of the crop to be harvested. Fourth, biotic diversity must be maintained above the level at which the activities of pests become a serious factor. Such diversity will reduce nutrient competition and avoid the depredations of pests.

The type of land use which most obviously satisfies these conditions is plantation farming of perennial shrubs and trees to produce rubber, resins, saps, oils, nuts, drugs and fruits. Some timber and forest products like charcoal, fuel-alcohol and silviplastics could be and already are being to some extent cropped and processed under strict management systems to ensure the maintenance of the overarching forest canopy and the steady replacement of trees at other levels.

On a different tack, the rivers of Amazonia, including the mighty Amazon itself, could support a prosperous fishing industry. This resource has been largely ignored to date.

All the products I have mentioned would be more likely to find a market outside Amazonia than the present cereals and other annual crops which can never be produced in quantities large enough to compete on national markets. These same commodities will always be produced more efficiently in other world areas with different soil structures, and much closer to large centers of population.

Another alternative to misdirected agricultural expansion in Amazonia is presented by Brazil's huge *cerrado* (grassland) regions to the south and east of Amazonia. *Cerrado* ecosystems are generally both more stable and more plastic to development than rain forest. The main reason *cerrado* has not been developed in preference to forest is the inaccessibility of the main *cerrado* areas, which range across sprawling highlands. A highway building scheme in *cerrado* regions would make much more sense, as a means of access to new farming lands, than the Amazonia highway project ever could. What is more, the kinds of farming techniques required to produce food from *cerrado* terrains are well within the capabilities of the farmers of the northeast.

Despite the existence of so many constructive alternatives, the Brazilian government is still totally committed to the construction of more and more highways across Amazonia. The emphasis of highway planning has, however, changed significantly since the agricultural limitations and health hazards of settlement in the forest region first became generally known. The *raison d'être* of the highways is now unmistakably the access they give to mineral deposits and to timber. The few forest regions suited to conventional agriculture are being rapidly monopolized by big-business farming concerns, mainly for the production of soy beans. The ordinary peasant farmers of the northeast now have very little place in this scheme of things. It is therefore impossible to see how the Amazonian superhighways can be justified in terms of their original brief, which was to help the peasant farmer. Instead, before this century is out, the world's largest remaining reserve of virgin tropical forest will almost certainly have been destroyed simply by and for cynical exploiters, including many western contractors who are household names to most of the readers of this book. This destruction is one of the prices the world pays for our standard of living.

That section of Amazonia which lies within Venezuelan territory and adds up to about one-fifth of Venezuela's total area is under similar threat. Despite the existence of tough-sounding legislation ostensibly designed to protect Venezuela's virgin forests from unmanaged exploitation, the impoverishment of this resource proceeds practically unhampered, especially in the area north of the Orinoco River. Venezuela is currently under pressure to increase her agricultural output and her access to mineral and other resources. A good deal of the great wealth Venezuela presently derives from her booming oil industry is spent on imported foods and fertilizers. When the oil runs out, local agriculture will have to supply needs now met by foreign capital, while new sources of energy for all forms of development will also have to be found. Yet lack of management and ecological knowhow characterize development projects in all the South American countries which abut on Amazonia. Such projects almost inevitably proceed at the expense of primary forest. In a very real sense, therefore, they can only set the scene for their own ultimate failure. Studies are in progress in Brazil, Venezuela and elsewhere to find ways in which development can be reconciled to environmental balance, but such studies are in their early infancy. It seems very unlikely that technologists will produce attractive or even any alternative development models before most of the damage is complete.

Much the same situation found in South America exists also in the rain and cloud forests of the Far East. T.C. Whitmore's(37) examination of the state of forests in Malaysia, Indonesia and Melanesia (an area which contains about 10 percent of the world's known plant species, nearly half of them unique to the area) leads him to the reasoned conclusion that: ". . . on present trends, primary rainforest here is doomed to disappear within our lifetime."

Comprehensive studies of the present condition of African tropical forests have yet to be made and published, but deforestation activities in Central Africa give even more cause for pessimism than in the Far East.

The issues raised by deforestation in its broadest sense are most clearly seen in the dry tropics, where the destruction of drought-adapted woodlands at the edges of the world's arid zones is a major factor in the spread of sand deserts.

Along the southern margin of the Sahara, for instance, the thorn-scrub savanna type of vegetation, dominated mainly by acacia trees, is being constantly reduced by shifting and settled cultivation, over-grazing, deliberately lit fires, and the felling of trees to provide wood for building and fuel.

In precolonial times, the vast semiarid territory ranging between present-day Sudan and the Republic of Niger was populated mainly by small groups of pastoral nomads. The flexible lifestyle of these individuals enabled them to take full advantage of abrupt seasonal and local changes in supplies of pasture and water, without placing a strain on the environment by remaining in any one place long enough to exhaust its resources permanently.

During colonial and postcolonial years, population pressures and moves to bring nomadic people under the supervision of central governments led to widespread sedentarization and settled farming. The good farming land was fenced off, leaving the nomads who preferred not to settle to seek more meager pastures in more arid places. The consequent extra vulnerability to drought then led herders to increase the size of their herds in an attempt to cut losses. Thus over-grazing, seen especially in the reduction of tree cover occasioned by the livestock browsing on seedling trees, is now a major cause of desertification in the region. But it is still by no means the only, or even the worst cause.

The commonest form of cultivation in the area immediately to the south of the Sahara is millet growing. Millet farms are created by clearing the ground completely of all trees, grass and herbs — usually by putting the area to fire and then hoeing it intensively. The European practice of leaving fields fallow for a season or two to allow the soil a period of recovery, incidentally, is virtually unknown.

No longer hampered by belts and islands of trees and long grass, winds dry out the exposed topsoil and rake it to a fine dust. Sometimes they heap it with drifting sand, so finally reducing its productivity to an insupportable level. After a dozen growing seasons at most, crop yields drop by about half and large areas of the spent land are abandoned altogether. Such recently "deserted" areas are plainly visible in satellite photographs as colorless zones separating the green areas of cultivation.

The scanty tree stock of the Sahel and other such areas is further threatened by the use of wood as fuel and as a building material. As an energy source there is often no alternative to wood (for cooking, smithing and so on) in semiarid tropical countries — while wooden huts and enclosures (the dwellings typical of such regions' settled areas) have to be frequently repaired or rebuilt because of insect damage and the testing extremes of climate. The average settled household's annual consumption of wood can total three hundred trees or shrubs. Extended, this figure signifies a loss of three hundred million trees annually from the zone between Niger and the Sudan(20).

In a few places (in parts of Mali, for instance) some attempt is made at replanting tree cover, but only on a very small scale in relation to the amount of deforestation. Moreover, much of the reforestation work that goes on in the Sahelian zone takes the form of plantations of economic trees (eucalyptus and other gum yielders, palms and so forth). There is little attempt to encourage the natural regeneration of native trees as part of agricultural development programs, either by limiting the effects of grazing and cultivation or creating stocks of trees for domestic consumption. As for keeping intact the woodlands needed as windbreaks, and as contributors to soil-restoration processes, the effects made are so piecemeal as to be almost meaningless. There is a brighter future in store for areas threatened by desertification if forestry in those areas can be supported, extended and diversified. If agriculturalists and rural development agencies can merge with forestry organizations to implement an integrated program, it will be brighter still. At present, however, these interests are, if anything, growing further and further apart. Specialized plantation growing of trees can never make up for the absence of natural woodlands in their proper settings.

WHAT ARE WE DOING?

It may surprise western readers to learn that they are heavily subsidizing the destruction of tropical rain forests in many parts of the world. Aid funds donated by western taxpayers are daily being spent to provide the tools of deforestation, the bulldozers and herbicides which spearhead the despoilation of these irreplaceable environments.

At the same time, funds subscribed by industrialized countries to international organizations — notably UNESCO — are being energetically applied to study programs, like UNESCO's "Man and the Biosphere" (MAB) program, which aim to call a halt to deforestation and other environmental dangers by increasing and publicizing knowledge of their damaging and irreversible effects. It is, on the face of it, a really insane state of affairs.

The MAB program organizers cooperate with, and report to, various national research and development organizations. As befits a UN agency, UNESCO policy is strictly apolitical and takes care to avoid any stigma of "aid imperialism." But, as a result, the studies it sponsors proceed only in districts and areas of research where national governments wish them to proceed. The end product of the research is a batch of suggestions and recommendations which have little political muscle.

UNESCO's research budget, though extremely large in total, is finely divided among numerous small projects, partly in order to involve as many interests and institutions as possible and so hopefully to boost the persuasiveness of research findings through combined influence. This policy, however, has a tendency to dissipate research efforts and to postpone the delivery of a comprehensive indictment of abuses like deforestation. One eminent physicist has likened UNESCO to ". . . an enormous herbivorous mammal, dispensing unnourishing milk from a plethora of tiny udders." Many a technologist will recognize how accurate this portrait is.

The unavoidable truth is that, in contrast to piecemeal and long-term studies, destruction of tropical rain forests must, for everybody's sake, be halted at least until we are fully aware of its full consequences and until every conceivable alternative has been properly explored.

International pressure on the governments concerned has to take a more concrete form than at present. There must be withdrawal of aid — or rather an increase of aid, on condition that it is spent on environmentally sound development projects toward the creation and maintenance of big rain forest reserves, or large-scale model schemes of forest development without deforestation. At present national parks and biological reserves cover only about one-quarter of 1 percent (0.25 percent) of Brazil's territory — alas, not a low figure by most Third World standards, but of course derisory.

International organizations have devoted as much attention to deforestation in arid regions as to rain forest deforestation. But the fruits of their researches again exercise little influence on policies actually and actively adopted by the governments concerned. One cannot help feeling that industrialized countries, for their part, are doing too little to use their influence to halt deforestation in the developing world.

I do not, by the way, mean to imply that aid-giving countries have a positive attitude toward trees in their own territory. They do not. Our disgrace is as great at home as it is abroad.

In fact, trees are generally undervalued throughout the entire world not only in respect of their ability to regulate the environment and provide timber, but as food and fodder crops. In a remarkable book *Forest Farming*, J. Sholto Douglas and A. de J. Hart (4) make a convincing case for agrosilviculture (the use of trees as an integral part of farming operations) as the food-producing technique best fitted to solve the world problem of hunger and to improve the natural environment at the same time. In his introduction to the work, Eugene Schumacher points out:

> Since fossil fuels, the mainstay of the "modern system," have ceased to be cheap and may soon cease to be plentiful, many people are becoming interested in solar energy. They are looking for all sorts of wonderful man-made contrivances to collect solar energy. I am not sure that they always appreciate the fact that a most marvellous, three-dimensional, incredibly efficient contrivance already exists, more wonderful than anything man can make — *the tree*. Agriculture collects solar energy two-dimensionally; but silviculture collects it three-dimensionally.

Less than 10 percent of the world's surface is currently used for food production. The rest is too cold, too barren, too dry or too inaccessible to support conventional crops. The great advantage of crop-yielding trees is that generally they can be grown in just such inhospitable places. Silviculture and agrosilviculture on a large scale could enormously extend the world's food-producing potential. In the view of Douglas and Hart, at least 75 percent of the world's present dry land could be made productive by growing trees either in managed plantations or in conjunction with less hardy crops.

Farm technology has long concentrated on mechanized, lowland agriculture. It is within this frame of reference that most of the innovations that have gone down in farming history have been made. Who knows what new techniques and processes might arise if technology switched its attention to the farming of trees and tree products? Certainly, there is no disputing the ability of trees to produce food, fibers, oils, fuels and plastics. But many highly productive and environmentally useful trees like the olive, almond, fig, locust bean, carob, mesquite, chestnut, pomegranate, oilnut, maple, kapok, tallow tree, datepalm and oak (to name only a very few) are gradually decreasing in numbers. On the other hand, annual plants like maize,

Table 5
Yield and nutrient content of some tree crops, compared with those of some annual crops.

Crop	Yield* under favorable conditions	Protein**	Carbohydrate**	Fat**	Calcium†	Iron†	Vit.A†	Vit.B†	Vit.B2†	Niacin†	Vit.C
Tree crops											
African locust tree (beans)	10-15	26	50	10	90	6.3	—	0.06	0.20	3.0	trace
Algaroba (pods)	15-20	17	65	2	260	4	—	—	—	—	—
Carob (beans)	18-20	21	66	1.5	130	3.8	—	—	—	—	—
Honey locust tree (pods)	15-20	16	60.5	7.5	200	3.8	30	0.33	0.13	0.9	—
Walnuts	10-15	16	15.5	64	99	3.2	—	0.32	0.38	1.2	—
Chestnuts	7-11	6.7	78.0	4.1	53	3.4	—	0.12	0.11	1.6	—
Dates	4-7	2.5	73	0.6	73	2.7	80	—	—	—	—
Annual crops											
Maize	2.5	10	71	4.5	12	2.5	—	0.35	0.13	2	—
Rice	0.5	7	80	0.5	5	1	—	0.06	0.03	1	—
Wheat	2.5	10	75	1	16	1.5	—	0.08	0.05	0.8	—
Groundnuts	5	15	12	25	30	1.5	—	0.50	0.10	10	10
Soya beans	3.2	35	20	18	200	7	—	0.30	0.30	2	—

*tons per acre (1 ton/acre = about 2511 kg/ha)

**g per 100g dry weight

†mg per 100g dry weight

rice, soy beans and so on are becoming increasingly familiar items in the landscape — even in areas where their culture is barely profitable and even though their yield could be greatly outmatched by, or powerfully combined with, that of productive trees (see Table 5).

The separation of agriculture from silviculture seems to have come about largely by reason of the different time scales involved and the economic consequences of that difference.

In the book *Forest Farming* a well-reasoned case is made for the multiple use of agricultural land, combining the culture of trees, pasture, cereal crops and so on, into one harmonious operation. The book also provides practical models for the design of "forest farms" — models which can be adapted to fit the special circumstances of no matter what environment. These authors emphasize that:

> The mixed forest is not a mere conglomeration of assorted plants, it is a highly complex system of checks and balances *adapted to the climatic and soil conditions of the area.* . . . An association of plants and animals is formed in given habitats under given environmental conditions. It includes plants synthesizing organic substances, animals feeding on these plants, carnivores and parasites living at their expense, and organisms capable of mineralizing organic substances that create conditions favorable for plants. . . . Numbers are maintained in the proportions *most advantageous to the area* by regulating mechanisms evolved during its historical development. [The italics are mine.]

Rain forest areas, temperate-zone farmlands, even semidesert, could all benefit from the three-dimensional approach. Food-production could also be improved vastly, without fear of ecological nemesis.

One factor above all others drives a wedge between these desiderata and their practical realization. It is the industrialization of agriculture in the developed world since the eighteenth century, a trend which has placed a premium on mechanized, large-scale farming operations producing the quickest and largest possible return for the smallest possible layout of capital and labor. Different kinds of agriculture (meat and dairy farming, silviculture, cereal farming, orchards and so on) have become segregated from one another in response to an imperative to simplify, to mechanize operations, and to keep labor forces down to a bare minimum.

The trades and industries which have developed to serve this modern norm — the fertilizer and pesticide plants, the heavy-machinery assembly lines, the brand-name food marketing corporations and so on — tie large numbers of people and incalculable amounts of public and private capital into the preservation of the factory-farm ideal. Developing countries in turn seek to imitate this model and are encouraged by developed countries through aid and private investment to impose it upon the same exploitation of their natural resources, however unsuited this approach may be to special local conditions.

It is hard now to step back and take stock of the basic efficiency and survival value of a factory system which has become so taken for granted and in which so many interests are vested. But that is what we must do, if we accept for example that the fossil-fuel, metal and cheap fertilizer reserves on which mechanized farming depend are not always going to be available. Forest farming, as envisaged by Frazer-Darling, Schumacher, Douglas and others, seems to be the ideal insurance against that inevitable day when the account books of conventional agriculture simply fail to balance.

Wealthier countries may, for a while, be able to buy and sell their way out of the problems arising from agricultural overspecialization. As long, for instance, as hardwood timber from the tropical forests remains a cheap alternative, there is no obvious need to develop a home-grown hardwood forestry. As long as pest and disease control, or fuel and machinery overheads do not reduce the profit margins of commercial softwood forestry to absolute zero in northern Europe, the USSR and North America, the amalgamation of forestry with other forms of agriculture remains (but for a few enthusiastic experiments) a mere notion.

The few relics of an earlier, saner alliance between farm and forest in industrialized countries, the small woodlands and hedgerows, are being rapidly erased by sweeping development or simply by neglect — as witness the lack of concerted efforts to preserve useful trees like the elm from virtual oblivion.

In poorer countries it is more obvious that the planting and replanting of trees is the best of few options open to farmers in their attempts to raise the productivity of their land and their own living standards. Schumacher(4) recalls that the Buddha:

> . . . included in his teaching the obligation of every good Buddhist that he should plant and see to the establishment of one tree at least every five years. As long as this was observed, the whole large area of India was covered with trees, free of dust, with plenty of water, plenty of shade, plenty of food and materials. Just imagine you could establish

an ideology which made it obligatory for every able-bodied person in India, man, woman and child, to do that little thing — to plant and see to the establishment of one tree a year, five years running. This, in a five-year period, would give you two thousand million established trees. Anyone can work it out on the back of an envelope that the economic value of such an enterprise, intelligently conducted, would be greater than anything which has ever been promised by any of India's five-year plans. It could be done without a penny of foreign aid; there is no problem of savings and investment. It would produce foodstuffs, fibers, building materials, shade, water, almost anything that man really needs.

Many industrialized countries simply do not have the space to enable them to become halfway self-sufficient in timber and other bulk forest products. But all the developed nations could do a great deal more than they do at present to maximize their stocks of trees and to accommodate the greatest possible diversity of tree species within their territory. And they should recognize that timber and pulp are not, except from the most blinkered point of view, all that trees are "for."

CONCLUSIONS

Big trees share with big whales one dubious distinction, apart from the fact that you cannot get either of them through the doors of a museum in one piece. They are, when alive, vital to the ecological balance of their surroundings and the survival of many other organisms. Nonetheless, they are cynically destroyed and exploited by humans for short-term gain.

An important difference is that trees, unlike whales, have not the faintest chance of escaping the destructive intentions of human beings, and this not only because they cannot get away. The hunting of blue whales has of late been restricted, albeit not very effectively, by international agreements on kill quotas. These agreements are based on the concept that blue whales are worth protecting because their influence on the global environment is too valuable to sacrifice to the trading interests of any single country. This same argument could be applied with equal, or greater, force to the predicament of forest trees, especially to tropical rain forests. But it never will be, for trees are firmly rooted to territory owned by this or that national government. The impulse to halt deforestation, in all its forms, and to reverse it has to spring up in the first place within national boundaries.

In the jungle of competing values that has grown up around our dealings with trees, one proposition must stand taller than the rest: we always need trees and there may always be times when trees need us. Any creature comfort we happen to reap from particular trees is a tiny bonus added to the inestimable gifts which the collectivity of trees has never so far failed to heap on us. A world without trees would be a world chasing its own tail, falling over itself to replace the irreplaceable. For all we know, the turning point between that world and the world we know pivots today on the stem of a single leaf.

7 Shadows on the Forest

"But here, instead of looking down on rich farmland, open and sunny, you look out over a vast sea of dense, shimmering tree-tops, so dark they are almost black. On all sides this ocean of forest stretches for as far as you can see, quiet and peaceful, asking only to be left undisturbed, or at least to be approached in peace. But the darkness of it all also carried a threat, a warning to outsiders that they would do well to stay away. The same darkness welcomes those who understand, and there is no feeling which is quite the same as the refreshing coolness of the first shadows cast by the leafy giants at the forest edge after years spent away from its shade and shelter."

COLIN TURNBULL
The Forest People (35)

THE WORD FOREST in its strictest modern sense means an area where trees and woody shrubs grow in the absence, or near absence, of grasses. The definition embraces many different vegetation types all over the world, but it excludes enormous areas which, though far from barren of trees, are ecologically dominated by grassy plants. It excludes, therefore, the various forms of wooded grassland found in the dry tropics north and south of the equatorial forest zone. Less obviously, it also excludes most of the parklike interweaving of woods and hedges characteristic of agricultural land in temperate regions — the prime domain of the elm, incidentally — and likewise the drought-resistant scrubland common not only all around the Mediterranean, but in many dry, rocky parts of warm, temperate and subtropical America.

Figure 10 shows an entirely imaginary land mass running from pole to pole, but having a distribution of desert and various forms of vegetation corresponding to those of present-day Europe-Africa.

Figure 10

This map of a hypothetical continent running the length of the 0° meridian shows the natural vegetation types which we could expect to find on any large land masses at a given latitude under present climatic conditions and assuming no outside interference (e.g. by man)

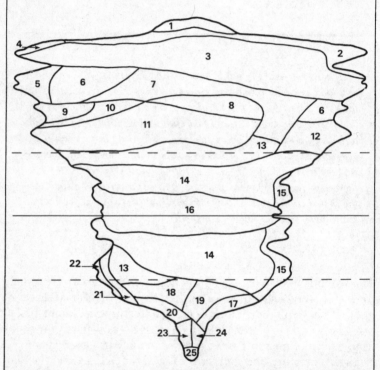

1. Arctic desert 2. Subarctic tundra 3. Boreal coniferous forests 4. Boreal birch woodland 5. Oceanic cold temperate deciduous and evergreen woodlands 6. Suboceanic cold temperate deciduous woodlands 7. Continental grass steppes 8. Temperate continental deserts 8. Temperate continental deserts 9. Warm temperate sclerophyllous woodland and shrubland 10. Subtropical winter-humid steppes 11. Hot deserts 12. Subtropical summer-hot monsoon-climates with evergreen broadleaved woodlands 13. Tropical dry and thorn savanna belt 14. Tropical moist savanna belt 15. Tropical rain forests based on orographic rainfall in winter 16. Equatorial rain forests 17. Subtropical rain forests 18. Subtropical thorn veld and succulent shrubland 19. Subtropical grasslands (Pampa, Veld) 20. Warm temperate sclerophyllous woodland and shrubland with summer dryness 21. Coastal deserts with moderately warm summers 22. Coastal deserts with "garua" 23. Cool temperate rain forests 24. Cold temperate steppes with mild winters 25. Subantarctic tussock grassland and moor.

The distribution of desert, scrub, forest and so on all over today's world at any given latitude is actually very like that shown here, given that the exact inventory of plant species differs from continent to continent. The different vegetation types they add up to are, however, close ecological equivalents to their counterparts elsewhere. There are slight but important quantitative variations in respect of Asia· and South America. These lands have (for the moment) a substantially greater amount of equatorial forest vegetation occupying their central lowlands. The question is, of course, whether we are going to continue to allow Asia and America to become more desert-filled like Europe-Africa. And equally importantly, are we going to allow the relative proportions of desert and scrub to increase everywhere on our planet, as we are presently doing?

Major vegetation zones can and should be as cosmopolitan as the climatic zones that lend them their identity. But it is important to realize that, like the climate, they are forever changing. It almost goes without saying that they change as a direct result of human activities. But they also change and evolve — as they always have — naturally.

The earth's position in relation to the sun and the sun's in relation to the universe can never remain absolutely constant. The climate therefore responds to cosmic changes and the vegetation has in turn to respond to the climate. Changes are also imposed from within the earth itself as land masses rise, fall or drift within the range of new climatic realities. Thus novel restraints and opportunities continually confront all plants and the animals they succor.

The earth's supply of water also pursues its checkered destiny, now overwhelming continents with shallow seas, now locked solid in ice caps that reach to the tropics. The soil is constantly washed away into the ocean and constantly reformed or reconditioned by the action of the weather. New soils, or newly depleted soils, cater to new vegetation types, and the vegetation, in its turn, further remodels the soil.

While all these changes are going on, uneven competition between different plants for basic but changeable (in quantity and quality) resources (earth, air, sun and water) leads to complex but predictable changes in the distribution and identity of this or that vegetation, and the catalog of creatures it harbors.

Ecologists regard the almost endless struggle that underlies such changes as succession. We have to say "almost endless" because if climate changes do not exceed fairly broad limits there is what is termed a "climax vegetation" in each major climate zone — a vegetation which enjoys such equilibrium between internal and external conditions that its character is stable and more or less self-maintaining. A climax vegetation is, so to speak, the finished product of natural succession. It is more complex, more independent of gifts of raw materials from its surroundings, more efficient at recycling resources, than any of the successional stages that precede and help perfect it. Any further changes imposed on climax vegetation from outside (dramatic climate changes, for instance, or man-modification) must almost inevitably lead to its replacement by less varied and less energy-frugal, although not necessarily less robust, types of vegetation. These are called postclimax or "dis-climax" successions.

A supreme example of a present-day climax vegetation is the tropical rain forest. UNESCO estimates suggest that tropical forests in general cover about two thousand million hectares of the globe — about half the world's total forest area. Some 42 percent of tropical forests are definable as rain forest. This type of vegetation grows in the moist tropics or subtropics at low or medium altitudes. It consists of evergreen, or partly evergreen, hardwood trees disposed in several (sometimes as many as six) distinct treetop layers. Each layer or stratum is subdivided into an assortment of tree species. Mature individuals of each forest species grow to a height characteristic of this or that stratum. The top stratum, or canopy, often grows up to two hundred feet high. A welter of animal life, especially of insects and worms, lives among the layers of the rain forest and in the soil beneath it.

This intensely concentrated and complex ecosystem requires massive quantities of nutrients to keep it going. Yet, as noted earlier, the soil of the forest floor is nowhere near as fertile as one might expect. The bulk of available energy is always currently tied up in the tissues of living plants and animals. It is the quicksilver recycling of dead organic matter that enables the system to thrive. Fresh, inorganic material (mineral salts, trace elements, etc.) is fed into the system, not from the ground as it is in most other environments, but almost entirely via rainfall and the infiltration of airborne dust particles. The soil is kept from clogging or being washed away by the dragnet effect of a huge number of plant roots and by perpetual reworking through the guts of soil-feeding animals.

In crudely hierarchical terms, the state just described is that toward which all "succession" in moist tropical lowlands strives. It is likely that there have been rain forests somewhere on the equator for millions of years, regardless of ice ages or ages of drought.

But virgin or "primary" rain forest, undisturbed by man, is now becoming a desperately rare phenomenon. Most modern rain forests, indeed most forests, are in "derived" successional states. These are the result of a mere dozen or so millennia of human exploitation and interference. Only the remaining virgin forests of the Amazon Basin now provide a truly large-scale example of a primeval vegetation completely unaltered by human activities. It seems almost certain, however, that they too will soon go the way of their more accessible counterparts in Africa and Asia, or perhaps even disappear completely within our own lifetimes. It takes centuries for disturbed tropical forests to regain their original complexity. Where such regrowth is allowed to proceed unhampered (as it very rarely is), it passes through a series of "secondary" successions, each more complex but more vulnerable to further exploitation than the last. Soil and water conditions may in the meantime have been so altered that succession, even when given free rein, stops short of full climax. As a result, the vegetation remains fixed virtually forever in a derived state.

Such was the ancient fate of most of the scrub terrains we now call chaparral, moorland, heathland and so on. Though travelogs and romantic novels portray them as the wild haunt of unsupervised nature, most of these wastes are, in reality, no more natural in origin than a parking lot or a cornfield. Centuries of overgrazing, overcultivation and burning led once upon a time to the irreversible impoverishment of the soil beneath them. The orderly circulation of water through that soil gave way to flooding and erosion in winter, then drought in summer. Where there was cultivation it had to be abandoned. Both wild and domestic herbivores competed for the leftovers until finally only a small number of rabbits and deer were able to live on the few remaining tough grasses and thorny shrubs.

It has to be strongly emphasized that virtually all these denuded areas used to be richly forested. It was this forest which mainly made the soil viable in the first place. In many cases, in fact, there is now no obvious climatic bar to the reforestation and eventual recovery of the land. Perhaps the most important limiting factor, apart of course from the intrinsic difficulty of raising forest plantations on exhausted soil, is — ironically enough — pressure from conservationists to keep these picturesque wildernesses at a rock-bottom, environment-bearing capacity.

In many parts of the world, deforested land meets a fate still more severe than conversion to moorland or scrub wilderness — desertification. Many of Arizona's drylands, for instance, were directly derived from woodland by straightforward burning. Early European colonists found a terrain still reasonably fertile and ready-made for large-scale cattle farming. It was intensive grazing, coupled with regular burning to suppress woodland regrowth, that certainly turned the original Arizona desert into a much larger stronghold of aridity than it could possibly have been without such help. When, recently, prohibitions were placed on deliberate burning in the grassland tracts fringing the desert, the consequent regrowth of thorny shrubs was viewed with alarm by ranchers as a threat to protein production. But these woody intruders were, after all, only growing where they naturally belonged. Left to their own devices, woodlands would eventually once again dominate the Arizonan landscape, constituting an extremely effective barrier to any further spread of desert. It is plainly in mankind's interest to interfere with natural succession in order to produce food. But in marginal situations such interference can so easily drive the character of the environment across the narrow divide between real estate and wasteland, and between wasteland and desert.

An even more dramatic example of the dangers is provided by the vegetational history of the Mediterranean and Saharan regions. It can be argued that maquis, the chaparrallike scrubland now typical of southern Europe's and North Africa's coasts and hillsides, used to stretch, as recently as five thousand years ago, right across the western half of what is now the Sahara Desert. Since then it has retreated into its present territory mainly as a result of man modification. Evidence exists to show that maquis itself, as we know it today, is a derived form of a yet richer vegetation, part of a long chain of successions of true forest vegetation which, in distant prehistory, may have flourished where now even the hardiest arid-zone plants peter out. Climate changes must have played an early part in the transformation, but certainly with a good deal of help from man after about 12,000 B.C. Human farming and herding activities, even without benefit of agricultural machinery or pressure from overpopulation problems of twentieth-century size, can permanently ruin the fertility of huge areas of land in a very short time, particularly where deforestation opens the account.

Today's maquis is also very different from pre-Bronze Age maquis. Pollen deposits found in past and present maquis regions testify that less drought-resistant versions, dominated by large laurel woodlands, once occupied those areas. These were supplanted only relatively recently by the dwarf oaks, junipers, heaths and so on which now give Mediterranean hillsides their distinctive character. The pre-maquis "laurisilva" vegetation now survives as a living fossil only in the moist, cloud-belted highlands of the Canary Islands and other eastern Atlantic archipelagos, and parts of South America. There are laurels to be found in modern maquis. But Canarian laurels are related to them far more distantly than they are to representative fossil species from the Sahara.

Since the period between the twelfth and fourteenth centuries A.D., when the European colonization of eastern Atlantic islands began, the vestiges of "laurisilva" have predictably lost ground, to be replaced by true Mediterranean maquis, unquestionably as a result of human exploitation. In the Cape Verde Islands, laurisilva has disappeared altogether, except on inaccessible mountain ledges. It has undergone, in the course of a few centuries, what the vegetation of the entire Mediterranean region underwent after the discovery of agriculture.

At first, seminomadic groups of people practiced shifting cultivation on a scale so minute that the rich native woodlands had ample chance to regrow between bouts of exploitation. As human nutrition and organization improved, populations became larger and pressures on natural vegetation became less redeemable. The unscheduled fallow period between the abandonment of worked-out clearings and their subsequent recultivation grew shorter. Need for building timber and wood also grew more and more pressing. The skill of metalmaking spread, requiring wood charcoal for ore smelting. Users of plows and other metal tools settled along soil-rich river valleys and coastal plains. Natural forests now became restricted mainly to mountainsides and other terrains meager in soil, badly drained or otherwise difficult to cultivate.

Later centuries saw a steady population increase around the shores of the Mediterranean, associated closely with the emergence of powerful city-building civilizations, notably those of ancient Greece and Rome. Wars of conquest added an influx of slave labor to the ranks of people working the soil. Rather than deprive their own agricultural peasantry of their role, Roman landowners set slaves to work deforesting and cultivating hillside and wetland sites of marginal fertility. The consequent soil erosion and flooding affected not just these badlands but also lands of prime productivity that happened to be linked to the same watersheds. Though soil conservation techniques (such as hillside terracing and artificial drainage) were well known to the Romans, their shortsighted priority was to exploit land to the full while it was plentiful and while labor was cheap or almost free.

Such an approach was not, of course, limited to the Roman world. It typified most precapitalist societies. By the time viable land became too scarce to treat in such a throwaway fashion, matters had often gone too far to reclaim the soil in all but the most favored sites. It became fit for recultivation only, if at all, after having lain fallow for centuries under a restorative cloak of forest. Prosperous agricultural societies in northern Europe of the Middle Ages owed their wealth to land that had been overworked and abandoned in pre-Christian times and had subsequently recovered a good deal of its fertility. Their use of this gift was more conservative and ingenious than hitherto, but still it stood at risk from the vicious circle that had strangled Roman agriculture. It still depended on the wholesale destruction of woodland.

Today, modern western agriculture, with its absolute emphasis on soil conservation, has made a fine art of escaping the same noose — but it has not made nations self-sufficient in food. The reason is that population growth long ago outpaced agricultural efficiency and even well-fed and highly developed societies, during the past five centuries, have looked to other shores (to the New World, then to the Third World) for extra raw materials (including wood), extra fertilizers and, above all, extra space. The greatest potential wealth of tropical developing nations lies not in the promise of their industrial output, not in their mineral or timber resources, but simply in the vast amounts of cultivable space harbored by their abundant natural vegetations. Here, as western aid policy explicitly imagines, are the world's future granaries.

But the exploitation of tropical land, whether it partakes of western funds and expertise or not, proceeds always at the expense of natural — especially forest — ecosystems. Yet the long-term environmental and economic consequences of the replacement of age-old forest and grassland realms by agro-ecosystems are still only barely guessed at. Local and worldwide population pressures apparently cannot wait for some fundamental assurance that agricultural development in the tropics is not prone to mistakes, just as irrevocable as those made by Bronze Age, Roman and medieval farmers.

When we consider the influence of man on forests, it is right that we should focus our attention upon present and recent man modification of tropical forests, for it is the most serious and urgent of all deforestation problems. It is still a problem we may be able to remedy technically — an ultimate test of our good intentions toward the environment we sometimes claim to chaperone. Yet we must not forget that humanity is not the only source of deforestation by a long chalk. When we look at the truly ancient history of tropical forests, we 'see that forests have been undergoing vast changes since long before man appeared on the scene, changes which continue today but on a scale and at a pace too gargantuan to be exactly appreciated, still less controlled.

The central theme of the history of land vegetation is an endless natural struggle between two major vegetation types — forest and grassland — for possession of the world's tenable dry surface.

Broadly speaking, grasslands form the climax vegetation in regions where annual rainfall averages are low or very low, and temperatures run to very hot or very cold extremes. Many natural grasslands shade into hot or cold desert or semidesert where these conditions are exaggerated. About a third of the earth's vegetated area is grassland going by the names of savanna, steppe, prairie, pampas, cerrado and many others. Grassland is not necessarily bare of trees. Indeed, some tropical savanna woodland is more like jungle than prairie, but the trees it supports are typical dry-country species and they plainly do not vie with the tall savanna grasses for dominance over the ecosystem. Thick-barked, thorny and slow-growing plants, they are actually too busy surviving to contribute much to the overall resources of the system. Wood-eating insects (certain termites, for instance) and a few leaf-eating or highly adapted large herbivores (such as giraffes) depend on dry-country trees for a regular supply of food, but grasses in such situations support a far greater tonnage of wildlife of all kinds.

Forest is normally found in humid regions, especially near the equator and in the cool, wet hinterlands of the temperate zones. True forest can be found in some dry regions. It can grow, for instance, in narrow strips along the courses of great dry-country rivers, or on the cloud-hung slopes of otherwise arid mountain ranges. Similarly, typical dry-country vegetation can survive along wind- and salt-dried coastal strips flanking large forest masses, and on the bleak ridges of mountains protruding from forest strongholds. Again, there can be "derived" grasslands or forests, created — if not altogether artificially — by aberrations in local weather or soil conditions. But normally forests and grasslands are able only to mix temporarily or locally, for their identity depends, as a rule, on the climate which prevails in this or that region. Quite slight but persistent overall changes in climate can cause revolutionary changes in the vegetation. Portères has, for example, argued that if the African climate as a whole altered toward conditions that were relatively humid, though with a long dry season, the entire continent would soon become, but for a narrow strip of forest near the exact line of the equator, an enormous prairie dominated by the savanna grass *Themeda triandra*(23).

Is there in fact any evidence that such an overthrow may indeed have happened in prehistory? Could it happen again, with or without extra pressure from human interference? Answers to these questions are found in the researches of paleobotanists and paleoclimatologists, whose interest it is to trace patterns of vegetation and climate through the geological record.

Though the history of Africa's vegetation should be more difficult to analyze than those of Asia or South America, where human influences have done rather less to confuse the picture, it is on Africa that controversy about ancient interactions between forest and grassland usually centers. There is, after all, a greater accumulation of information about the present-day distribution of African plants, owing to a longer history of western colonial influence and the cataloging urge of amateur and professional botanists. From such information, deductions about the primeval distribution of plants can be drawn. Such theorizing has already been rife for well over a century, and today it draws new strength from modern methods of biological interpretation and fresh techniques of obtaining and using fossil evidence.

A modern vegetation map of Africa shows a massive bloc of forest straddling the equator and extending some way into south-central and southeast Africa. To the north, savanna zones, each more arid than the last, open toward the sand desert of the Sahara like the pages of a book, clearly reflecting sharp reductions in the average length of the wet season in relation to that of the dry season. These divisions also relate closely, as one might expect, to the pattern of isotherms (lines of constant air temperature) across the continent.

Apart from some neighborhoods near giant mountain ranges, or very close to the equator, most tropical areas have a Jekyll-and-Hyde climate which alternates between wet and dry seasons in abrupt succession. The tropical dry season runs, very roughly speaking, concurrently with winter and summer in Europe and North America. The wet season corresponds to our summer and early autumn. The relative length of the wet and dry seasons determines the climatic regime of any tropical locality. But that regime does not always rule evenly. Often the wet or dry season will last a shorter or longer time than usual. If such an aberration is repeated during several successive years, its status alters. It ceases to be a long-term weather change and begins to become a short-term climate change. As such, it need only last for a handful of years to work spectacular changes in the vegetation of the locality affected. To give a recent instance, two abnormally dry summers in Britain (1975 and 1976) led to the death of millions of trees in a mere three-month period — April to June 1976. Predictably, fire added its own contribution to the problem. The number of acres of woodland destroyed by fire during that same twelve weeks was twice the annual average for other years. For a while, much of the open lowland countryside within fifty miles of London had a parched and bare appearance not at all unlike that of African savanna woodland. Transfer this scene to the dry tropics where zero rainfall for three months of the year is commonplace, and it is quite easy to see how woodland can become grassland, and grassland desert, in an extremely short time, even if the climate deviation persists for just a few years, as happened recently in Africa's Sahel savanna zone.

When, on the other hand, abnormally moist conditions shorten the dry season, fast-growing tropical forests and woodlands soon expand into the previously open territory. Conditions may subsequently return to normal, but these pioneer stands of trees will perpetuate themselves, at least for some time, once firmly established. Climate fluctuations can thus have a flywheel effect on the vegetation, manifesting themselves across a timelag that makes their overall impact, however visible, nevertheless a complex affair. Even the grandest climate changes, such as those which mark the beginning and end of interpluvial periods and affect tropical continents in their entirety, do not, so to speak, sweep the field without further ado. They, too, are tempered by temporary or local fluctuations against the general climatic trend and there are always a few spots — wet or dry — which will retain the once-characteristic vegetation regardless of the degree of change experienced elsewhere.

In Haffer's view(8), a particular type of vegetation could, year by year or through alternate series of years, appear either as an uninterrupted cover or as a mosaic of isolated patches, under the influence of staggered sequences of climatic ebb and flow. Haffer reached these conclusions on the basis of studies of the tropical rain forests of the Amazon Basin, but his views are equally applicable to the vegetation of Africa or Asia. All tropical forests have been reduced, during long periods of low rainfall, to small numbers of bridgeheads from which subsequently, under wetter conditions, a new forest offensive may be mounted. Haffer envisages the South American equatorial forests — the largest remaining uninterrupted treescape in the modern world — reduced to nine small, isolated refuges under the kind of hot, dry conditions which were very probably attained at the height of the last interpluvial period. Haffer goes further to argue that, since the present day is probably part of the early stages of an interpluvial period: ". . . the present continuity of the Amazonian forest seems to be a rather recent and temporary stage in the vegetational history of South America. . . ."

If equatorial forests are perhaps only "recent and temporary" visitations, their fate may well nevertheless affect that of the human race. If a natural overall fluctuation of tropical climates in favor of dry conditions is reinforced by "disease deforestation" in temperate lands and by deliberate manmodification and destruction of forests worldwide, there may well soon be real substance in the disturbing image of a world without trees. It is quite possible that deforestation within the next fifty years will measurably modify the world's climate and atmosphere, and will have begun — in this and many other ways — to limit processes of human development which have for long relied on both the presence and the expendability of trees. The very survival of the human race may be placed at risk.

There is nothing particularly fanciful about these suggestions, but there is, as we have repeatedly seen, a depressing lack of means by which to measure the seriousness of the problem and place it beyond dispute.

What forms does man modification of forests take in today's tropics? Replacement of forests by simplified agricultural ecosystems is the largest major category of deforesting activities. Shifting cultivation, though practiced without machinery and usually by small-scale societies, yet amounts to the most irresistible of such activities.

Shifting agriculture is the way of life of an estimated two hundred million people and affects, at any one time, about fourteen million square miles of the globe's surface. In other words, about 5 percent of the world's human population farm about a quarter of the world's dry land (or half its cultivable land) in this fashion. Shifting cultivation is most common in the forested humid tropics, where the people who depend on it for their survival often (though not always) have nomadic or seminomadic lifestyles. The "slash-and-burn" farmers create space for food production by leveling a patch of natural vegetation and removing most of the tree roots. When the productivity of cleared ground fails, as it soon must without support, the farmer makes a new clearing elsewhere or returns to an old clearing previously exhausted but restored through disuse to a measure of its original fertility.

Shifting cultivation does not inevitably cause permanent damage to natural ecosystems. While it remains the occupation of small societies lacking factory-made tools, it can harmonize with the environment to an extent that continuously managed agriculture cannot possibly match. For, in this form, the original vegetation is allowed plenty of scope to recover lost ground, having temporarily ceded a tiny fraction of its resources to the farmer. But recovery in this context must be measured in decades or even centuries, and it is quite clear that human population pressures have increasingly reduced the period of rotation of clearings to fewer and fewer years. Loss of soil structure followed by soil runoff and invasion by forest-preventing weeds are the predictable results.

Certain parts of tropical West Africa currently bear graphic and tragic witness to the long-term effects of overintensive shifting cultivation. There, once-vast woodlands have commonly been reduced to narrow roadside thickets — which give the traveler an illusion of an endless treescape. Leaving the road, however, he soon finds himself in a wasteland of abandoned cultivations, some of which have been infertile for two centuries or more since they were last cleared.

All tropical woodlands and forests are liable to meet the same fate. The only reason certain tropical forests have so far escaped it is because they are rather inaccessible to humans other than a few small societies whose ecology harmonizes well with that of the tree communities which nourish them.

A wide range of variant forms of shifting cultivation can be identified. This range includes the use of modern herbicides as a clearing tool, the practice of leaving large trees standing while clearing only undergrowth and lower-story trees, and so forth. Beyond a certain level of sophistication such practices outrun the traditional definition of shifting cultivation. Their side effects are, even so, no less deadly than those of intensive shifting cultivation.

Timber exploitation probably comes second to shifting cultivation as a cause of permanent deforestation in the tropics. Sometimes this goes on in conjunction with agricultural programs, but in its most common form it simply involves the clearing of large areas of forest for the sake of access to a small number of desirable timber trees.

"Worthless" trees, the stripped foliage of timber trees and undergrowth, can account for 60 percent of the weight of plant life destroyed in the process of felling operations. The usual fate of the cleared area is to bear cultures of annual crops for a few years until soil fertility is lost or weed and pest problems succeed in making farming uneconomic.

Very occasionally, clearance is combined with forest management procedures. The "tropical shelterwood" and the "Malayan uniform" systems are examples of such procedures. Their aim is to control natural forest regrowth so that offspring of valuable timber trees are encouraged to grow at the expense of unwanted trees and undergrowth. This aim is usually achieved by selectively poisoning the latter, either before or during the initial exploitation of mature trees, and by controlling subsequent weeds until the success of the new stock is assured.

Sometimes artificial regeneration of timber-worthy forest is achieved — or natural regeneration supplemented — by planting nursery-reared lines of desirable trees along totally cleared strips. Monocultures of fast-growing pines, eucalyptus and other exotics are increasingly preferred to mixtures of indigenous tropical hardwoods for this purpose.

In some parts of the tropics, notably in Burma and West Africa, a management system known as *taungya* is used. Seedlings of indigenous trees are planted in widely spaced patterns on cleared land shared with agricultural crops. The plots of arable land are allocated to local farmers in return for their undertaking to care for the trees until they are well established. For at least three years, the farmers tend crops between the growing trees. When trees and crops begin to compete seriously with one another for light and other resources, the farmer who had satisfied the terms of the original allocation, is offered a new plot on the same terms. A very similar practice, under the name of "*shamba*," is followed in parts of East Africa, but there the trees used are usually representatives of softwood species, grown in monoculture.

Most of these management practices are derived from techniques originally developed in the West under very different conditions. With the possible exception of *taungya*, none of them has been uniformly successful in the tropics and none of them is very cost-effective by comparison with large, managed forestry enterprises in North America or Scandinavia. Most of the world trade in timber is accounted for by softwood. Hardwoods are reckoned to command less than 25% of the market. And the very features of tropical hardwood forests that so excite the interest of the ecologist — their complexity, their luxuriance, the interdependence of their diverse components — are nothing but obstacles to the commercial timber grower. His ideal crop is one of standard workability, standard size and amenability to mechanized harvesting and handling. Tropical forests can be replaced by softwood plantations, but only against formidable opposition from weeds and pests and at the risk of failure through loss of soil fertility.

Another and more common form of forest replacement is the plantation farming of nontimber tree crops such as rubber, cocoa, palm oil and so on. Plantations can be founded by means of agrosilvicultural methods like *taungya*, coupled with high capital investment in fertilizers and pesticides.

Sometimes again forest can be replaced by plantations of banana or other perennial nonwoody crop plants. This use, like the others mentioned, can be planned as one stage in a strategy that permits forest regrowth during long fallow periods. But seldom, except by chance, is this technique adopted across large areas.

The capacity of deforested land to support crops usually stretches only to the temporary accommodation of tough annual cereal and root crops, such as maize or cassava. Even when cultivation programs form part of larger scale agricultural programs, partaking of all the benefits of modern farming — machinery, fertilizers and so on — the fine structure of the ex-forest soil all too soon loses its power to marshal water and nutrients. Repeated parching and flooding disables the new land after a few profitable harvests. It becomes a victim of what is only a mammoth version of shifting cultivation.

Much the same story goes also for forest tracts converted to grazing land, though thanks to the lesser degree of soil disturbance involved, there is often a slightly better chance of forest regrowth on abandoned pastures than there is on abandoned plowlands.

Forests are also, these days, often replaced by urban landscapes. New industry and settlements in the tropics are frequently sited in forests as part of strategies to create centers of population in areas considered to be underused.

Needless to say, the Amazonian or the Zairean rain forest is by no means empty of humankind. It is the home of hundreds of small-scale cilivizations. The people who belong to these forest societies follow a great variety of lifestyles, each finely adapted to make the most of unique surroundings. These styles range from subsistence on wild goods like honey, nuts, eggs, insect grubs, fruit and so on, to hunting and limited herding and shifting cultivation. Most forest tribes live by a flexible permutation of all these means.

When large-scale industrial societies impose their technology and values on forest-adapted societies, the latter lose their identity virtually overnight. In the new towns and around the new highways, people who once followed an independent and expert existence deep in the forest now surface as pathetic hangers-on to the utterly alien communities which have suddenly appeared in their midst. They now gravitate to rural slums where they live in abject poverty, a prey to epidemics of imported diseases to which they have low, or no immunity. Their language and social systems all but disappear in a very short time after their first contact with "civilization."

The effect of large deforestation programs on settled peasant communities, as distinct from nomadic tribes, is less dramatic but in the long run scarcely less harmful. The rush for crops of timber makes much extra land available for farming and creates a temporary, false boom condition. Standards of living and birth rates rise only to slump again a few years later. The community must now either pull up its roots and move to a new site or become poorer than it was at the start. Then it farmed far less land but did not thereby pose any threat to the stability of the forest soil. So no real economic progress is made. The problem is prolonged *ad infinitum* or soon brought to a gloomy crisis.

In Mexico, where deforestation has whittled the last of the country's large forests down to two million hectares, government technologists have at last recognized that "... neither temporary timber exploitation nor the cultivation of corn result in stable or fair economic or social conditions"(36).

The failure of soil fertility is by no means the only negative consequence of deforestation. For a start, the incidence of agricultural pests and diseases is often greatly increased. There is a well-established school of thought which maintains that the replacement of forest by simplified ecosystems inevitably leads to severe pest outbreaks, on account of the reduction of biological diversity. Such loss of diversity actually means at least three different losses. Ecosystem diversity, or the richness of the pattern of different vegetation types across a large region, is the thing most obviously lost. Life form diversity, or the variety of different major types of organisms (e.g., birds of prey, carnivorous mammals, marsupials, forest vines and so on) able to maintain habitats in the neighborhood, is also highly liable to disruption. Finally, taxonomic diversity, or the number of genetically separate populations (species) of one life form or another to be found within the ecosystem, is most seriously — though least conspicuously — diminished.

The reasons biological diversity is considered to have a restraining effect on the incidence of pest outbreaks are discussed elsewhere (see p.133). Broadly speaking, if you reduce the complexity of a tropical forest by cutting down timber trees and the various undergrowth plants, you instantly reduce the number of ecological niches that actual or potential pest- and disease-control agents can occupy, and you also greatly impair the entire ecosystem's resilience — its ability to grow back to normal after a serious disturbance such as a pest outbreak. Sometimes the agro-ecosystem that is established on cleared ground may be afflicted by infestations of animals that were not pests at all until clearing forced them out of their natural habitat.

Conversely, the more complex or diverse an agro-ecosystem, the less prone it usually is to pest or weed outbreaks. Crop monoculture (large-scale farming of a single crop) is from this viewpoint held to be the most risky of agricultural procedures with which to follow deforestation, however economical it may be in terms of standardization of tools, techniques and marketing. The expense often entailed in pest and weed control can easily make the enterprise thoroughly uneconomic.

Nor are the pests and diseases liberated by deforestation confined to plants and cattle. Human diseases are just as liable to escape natural control. Recent serious outbreaks of malaria in Southeast Asia are known to have begun in this way. A forest species of mosquito not previously considered to be a pest was introduced into human settlements in areas of cleared forest by force of the new circumstances. It became a disease carrier which, since it had considerable resistance to the insecticides effective against recognized pest species, soon gained a range of distribution of thousands of square miles. Formerly it had been confined to a few tiny pockets in one small area of rain forest.

The eleventh-hour recognition of these and other drawbacks to conventional methods of exploiting forests has led many tropical states to seek new methods of putting them to use. These are ways that do not necessarily involve destroying the forests at all — or which at least limit the waste of human and natural resources. They even offer some hope that a twist of events could make reforestation a commercial proposition.

It is not surprising that Mexico is one of the countries which has taken a lead in research to find novel uses for plants growing naturally in forests. A large part of Mexico's export income is earned by trading in a raw material for birth control drugs, derived from a vine which forms part of a rain forest ecosystem. New foods, drugs, textiles and pesticides certainly await discovery in the world's tropical forests. If their identity and value can be pinpointed in time, perhaps the future of the forest can be assured for as long as the supply of and demand for these commodities can be made to last. The race to find these resources is, alas, greatly impeded by the disintegration of small-scale forest-dwelling societies that has taken place in recent time. In the medical practice of those societies lie many clues to the potential of forest plants and animals in enhancing human life. It is a sad irony that the existence of "savage" cultures is so often represented as a hindrance to economic development, when it might, on the contrary, provide us with reconnaissance that could create a better life for everybody. It was, incidentally, to the customs of local tribes that the obscure discoverer of a marketable source of birth control drugs directed his search.

If forest exploitation cannot be achieved without deforestation, at least it can be achieved without senseless waste and the total destruction of the environment. This is the lesson Brazil has partly taught by using sugar-cane waste and the byproducts of timber exploitation to manufacture fuel-alcohol. Their example deserves imitation, for it means that the debris created by clearance is put to a good use rather than being burned or otherwise piled into the atmosphere as surplus carbon dioxide. What is much more, if practiced on a large enough scale, this process may well prove profitable enough to justify the nurture of new forest to ensure future yields. Perhaps inevitably some diversity will be sacrificed in the process, and timber management systems be imposed during the fallow period. This technique of producing liquid fuel from dead foliage is still experimental. But it seems, on paper, capable of making Brazil self-sufficient in fuel by the end of the present century.

So alternatives do exist to the permanent destruction or total replacement of forests, ways to conserve them that are not incompatible with commercial developments. Broader arguments in favor of conservation reinforce the desirability of such tactics. At a national or local level, watershed management and the usefulness of the remaining unspoiled forests as recreation areas or tourist attractions are the most concrete of these arguments.

At international levels there is also plenty to be said for safeguarding the future of tropical forests for the sake of their effects on the global balance of atmospheric gases (see Chapter 2) and their importance as gene banks.

The number of different kinds of organism the earth can accommodate is enormous but finite. A much greater (though still finite) number of genes present in the chromosomes in each species' cell nuclei promote the differences between organisms. Genes can recombine in novel ways and manifest themselves in new (hybrid or mutant) organisms. But such innovations are rare in the wild for various reasons connected with the physical limitations of the genes themselves, and with the fact that the odds are heavily against a new organism finding an ecological niche it can occupy or successfully compete for.

Each time a species of animal or plant becomes extinct, the world loses some genes which can never be replaced or reproduced. Each gene, if it happens to continue to exist, has the potential to make a unique contribution to the way the living world functions. If genetic engineering (the artificial mingling or creating of genes) ever becomes a safe and feasible technology, all the world's existing genes will be valued individually as possible raw materials for the improvement of our own lifestyles and our influence on the world about us. Today, in the more humdrum world of hothouse experimentation to find better breeds of crops, the search for useful genes has become urgent and ambitious.

The greatest variety of raw-material genes (and, it follows, of species) is found in undisturbed natural environments in the tropics — especially in rain forest. Such environments are therefore often referred to as gene banks.

A powerfully convincing example of the need for gene banks is provided by the events which followed recent worldwide failures of potato crops due to outbreaks of fungal disease. Frantic breeding experiments failed time after time to produce a disease-resistant variety of this staple food crop. Finally, a search was made in the Peruvian Andes — the natural habitat of the wild potato plant. Since it was first popularized as a food several centuries ago, the potato had been selectively bred to increase its yield per acre. In this the breeders succeeded magnificently. But somewhere along the line the potato lost the gene or genes which bestowed natural resistance to certain fungal diseases. Only by a visit to the gene bank of the Andes was that resistance recovered and reintroduced into the general heredity of seed-potato stock.

The relevance of the billions of genes incorporated into tropical forest ecosystems to actual human or environmental problems is not possible to judge at present. The size of the task of cataloging the external characteristics of even very conspicuous forest life forms cannot be conveyed without seeming to exaggerate. In one 25-acre patch of Malaysian rain forest, 227 different species of tall tree were identified. Who can say what multiple of 227 all their parasites, symbionts and host-specific foliage-, bark-, root- and sap-feeders might amount to? Not to mention the predators, the predators' parasites and so on. To say nothing of the hundreds of soil- or litter-feeding invertebrates that cram the forest floor!

It now seems that we shall never realize the full potential of the tropical forest gene banks. For there is no hope, despite the attractions of the alternative policies described, that the tropical forests, as they now stand or recently stood, will be widely conserved. A few small reserves have been marked out. But the size and number of these are nowhere near great enough to provide, for example, an ultimate guarantee that tropical forests could, if necessary, be recreated on anything like their former scale. This much we have lost already. In addition, the significance of these inimitable ecosystems to the global environment will soon have been tested to destruction. Perhaps — if we are incredibly lucky — it will be discovered that their loss does not pose insoluble problems. Perhaps on the other hand — as seems likely — the environmental impact of their complete destruction will show our present treatment of them to be the gravest human folly of all time.

In the absence of concerted reassurance from technologists on this score, it is surely time that the public at large came to decide for itself whether or not the risk is acceptable. We in the West cannot, beyond a certain point, preach different conservation or land-use policies to Third World governments. Their economic predicament is often such that they cannot prevent or forego deforestation, any more than western governments can prevent or forego the use of agricultural fertilizers and pesticides. But we who live in aid-giving countries can make sure that aid spent on development projects which currently involve sweeping deforestation is given more conditionally than at present. We can press for the creation of more, and bigger, tropical forest reserves — the cost of creating and maintaining these to be borne internationally. We can insist that this happens.

Most important of all, we ourselves must eschew armchair conservation and turn our attention to the ways in which forests and woodlands are exploited and destroyed in our own countries. We can look for ways in which they can be protected and developed so that they contribute something more than sawdust and cellophane to the wellbeing of humanity, and to the environment by courtesy of which humanity exists.

8 Death of Hedgerows

"Thus were the visions of mine head in my bed; I saw, and behold a tree in the midst of the earth, and the height thereof was great.

The tree grew, and was strong, and the height thereof reached unto heaven, and the sight thereof unto the end of all the earth.

The leaves thereof were fair, and the fruit thereof much, and in it was meat for all: the beasts of the field had shadow under it, and the fowls of the heaven dwelt in the boughs thereof, and all flesh was fed of it."

DANIEL 4:10-12

THE LOSS OF hedgerows in Britain peaked in the 1960s with an average destruction of six thousand miles a year. This peaking was due not so much to a return to sanity by farmers and government agencies, as to the fact that less hedgerow was, and is, now left to destroy. Since the end of World War II up to the present, the total aggregate loss averages out at five thousand miles for every year.

The plight of French hedgerows is no less serious. *Remembrement* is the fashionable euphemism for the systematic destruction of hedgerows, stonewall boundaries and small woodlands in several major food-producing areas of France, with the object of facilitating mechanized agriculture. This government-sponsored program began more than two decades ago and has since taken an average annual toll of four hundred thousand hectares of tree cover at a cost approaching nine million dollars a year.

Following the first implementation of the scheme, two years of intensive dehedging had run their course before a word of advice was sought from ecologists or environmentalists and only in 1978 was a thorough study of the side effects of *remembrement* published — a study which itself cost about two million dollars to prepare. And the chief conclusion drawn by the authors of the study report is that *re-membrement* has, to date, posed as many problems as it has solved and should proceed, if at all, with the greatest caution and a far greater regard for conservation than hitherto.

Ten million hectares of old boundary hedges, ditches and walls have so far been cleared; a further eight million hectares are scheduled for disposal over the next twenty years.

Nowhere is the evidence of the folly of the scheme so abundant as in Brittany — one of France's main livestock-farming areas. The chief argument in favor of *remembrement* in this area was a need to plant maize monocultures on a large scale so as to supply a ready-to-hand source of poultry fodder. The drastic remodeling of the Brittany countryside for this and other, more general reasons brought some advantages for some Bretons: farming was, of course, made easier thanks to mechanization; the exodus of young people from rural to urban areas was interrupted to some extent; the real-estate value of smallholdings on remodeled land rose sharply. But crop yields did not, on the whole, improve as convincingly as had been predicted, while in several parts of Brittany *remembrement* cost dear to those who suffered by the erosion of soil and by sudden danger in the form of serious flooding.

In 1974 and again in 1977, the town of Morlaix suffered floods worse than any known in the area for a century. Quimper and Chateaulin experienced similar disasters, with hundreds of thousands of dollars' worth of damage caused. Government scientists admitted that the removal of natural obstacles to flooding (hedges, ditches and so on) and the rerouting of streams and drainage canals as part of *remembrement* activities, were a direct cause of the trouble. In many areas, extensive replanting of hedges (a process far more costly than dehedging) has been necessary to stem floods and prevent further erosion.

Social problems also reared their heads in this normally tranquil and easygoing part of the world. Long-forgotten ownership and boundary feuds were revived in deadly earnest. Age-old systems of land tenure were challenged for the first time since the revolution. Doctors and social workers reported an abnormal increase in cases of depression and suicide in neighborhoods where *remembrement* was in progress.

The more fiercely independent and traditionalist of the Breton landowners organized activist groups to resist any further extension of the program in their area. Finistère, in western Brittany, became the focal point of such resistance and *remembrement* was actually discontinued there in 1972 as an act of appeasement. Recently, however, destruction has begun anew and thirty thousand kilometers of hedgerow (equivalent, of course, to a forest thirty thousand hectares in area) has already disappeared during eighteen months of *remembrement* in just two neighborhoods of Finistere.

The death of hedgerows is intimately linked with the death of trees. Both are intimately linked with the death of soils and the death of wildlife. Writing on the mass destruction by poison spray of millions of miles of wayside vegetation in America (the equivalent of the British hedgerow), Rachel Carson(2) noted:

> Of some seventy species of shrubs and vines that are typical roadside species in the eastern states alone, about sixty-five are important to wildlife as food. Such vegetation is also the habitat of wild bees and other pollinating insects. Man is more dependent on these wild pollinators than he usually realizes. . . . Some agricultural crops are partly or wholly dependent on the services of the native pollinating insects. Many herbs, shrubs, and trees of forest and range depend on native insects for their reproduction . . . without these both wild animals and range stock would find little food.

In their pamphlet *Hedges*, the Council for the Protection of Rural England notes that a high proportion of the fauna and flora (animals and plants) of the countryside depend absolutely on hedgerows. Forty species of birds are noted in this connection — including thrush, chaffinch, linnet, blackbird and yellowhammer. Some species, like the partridge, are already severely depleted. About two hundred and fifty flowering plants and ferns also survive mainly in hedgerows. Some seventy of these already face immediate extinction as matters stand at present. If current rates of "progress" continue, the remainder will shortly and surely follow.

In fact, few wild plants and animals can survive on arable land alone. For the wild animal in particular, the hedgerow is not simply a home. The continuous, interlinked network of hedgerows functions as a corridor. If wild creatures are unable to move freely from one place to another, they cannot survive in balance.

Hedgerows, like trees and woodland, also function as windbreaks, thus preventing soil evaporation, and as barriers to soil erosion, and serve to reduce the incidence and extent of flooding.

Even with that said we have not exhausted the direct potential value of hedgerows to farmers.

In their book *Forest Farming*(4), J.S. Douglas and A. de J. Hart write:

> In Britain the possible role of shelterbelts as a source of food appears not to have been considered, but the Russians recommend pear trees and crab-apples as suitable for shelterbelts. . . . Wild cherries and hazels can also be planted, and there is no reason why blackcurrants and cultivated blackberries should not be included in the shrub layers, provided they are well fenced against stock.

The authors further note:

> The potential of the traditional English hedgerow as a source of food for both human beings and livestock seems to have escaped most modern farmers, and yet such typical hedgerow plants as the elder, the hazel, the wild rose, the willow, the beech, the ash, the elm, the alder and the oak are rich in minerals and trace elements, which their deep roots draw from the subsoil, and are greedily browsed by animals, even in the depths of winter. Those farmers who have in recent years bulldozed thousands of miles of hedgerows . . . will surely come to realize that they have deprived themselves of a resource which could have gone a long way towards meeting the increased cost of imported feedstuffs.

The death of the British hedgerow and of American roadside vegetation, just like the death of trees, is entirely symptomatic of the modern two-dimensional approach to farming — the approach which considers only the surface area of cultivable soil. That approach, apart from its other failings, totally neglects the potential of the space above the surface — the third dimension into which hedges and trees reach. In an optimum rural economy, that third dimension would add many millions of acres of cultivable surface to our natural resources; a surface suspended, literally, in the air.

Such considerations aside, the tree, the hedgerow, the hedge plant, hedge animals and insects are the creators and maintainers of the soil. The farmer's crop draws on the bounty they create and protect, a bounty currently still in surplus, despite persistent exploitation and misuse. How could we possibly be better off without it?

9 Who Needs Wood?

*Riven deep by the sharp tongues of the axes, there in the redwood
    forest dense,*
I heard the mighty tree its death-chant chanting. . . .
Murmuring out of its myriad leaves,
Down from its lofty top, rising three hundred feet high,
Out of its stalwart trunk and limbs, out of its foot-thick bark,
*That chant of the seasons and time, chant not of the past only but the
    future.*

WALT WHITMAN
"Song of the Redwood Tree"

DEAD TREES ARE in demand. The USA alone consumes more than
enough of them each year to build a twelve-foot boardwalk to the
moon. Several European countries rely for their supply on imports
only slightly less expensive than their oil or uranium requirements.
Ninety percent of these imports, we note, are softwoods — princi-
pally spruce, larch and pine.

Table 6

*Estimated world consumption of wood in 1975, classified by
types of tree.*

	Millions of cubic meters	Percentage
Softwoods	1120	44.8
Tropical hardwoods	950	38.0
Temperate hardwoods	430	17.2

Even so, less than a tenth of the annual world crop of two and a half billion cubic meters of wood actually leaves its country of origin. The large bulk of the crop goes for home consumption. Of this almost half is simply burned, either as fuel or as waste in clearance programs. The remainder forms an essential raw material resource for a wide range of industries. Significantly, nearly all managed production and consumption of timber takes place in the industrialized countries — in the USA, the USSR, Canada, the EEC and Japan.

Probably the most familiar user of timber as such is the building industry. A two-bedroom suburban house can easily incorporate half a dozen mature forest trees in its roof, doors, windows and flooring. Although it is true that the proportion of wood used in individual buildings — especially in the high-rise block — has itself declined due to the increased use of synthetic materials, the constant growth of human population sees to it that the market requirements of the construction industry increase steadily.

Less buoyant is demand in the furniture industry, where synthetic materials can compete more successfully for most purposes. It is worth emphasizing that it is the rising cost of wood itself as much as any other factor which induces substitution.

The amount of timber used both in construction and allied industries is, however, far outstripped in developed countries by that used as pulp for papermaking.

When Aldous Huxley was writing his analytic essay, "Writers and Readers"(11), in 1936, he could confidently cite that the output of printed paper in Europe and the USA was of the order of ten million tons. In those days a combined total of forty thousand new books were published every year in English, French and German. Today thirty thousand new titles are published annually in Great Britain alone. And the North American figure is too high to count. So great has become the flood of printed material of all kinds, especially newspapers, that a realistic estimate of the world consumption of paper for printing can no longer be attempted. We could, however, multiply Huxley's figure by ten without risking exaggeration.

It is true that paper is also sometimes made from other materials — such as cotton and linen rag, or esparto grass — but trees continue to provide the very large bulk of press fodder. General economic factors act to ensure this continued use of timber for almost all papermaking. Trees can be supplied in virtually constant sizes and weights and they are easily transported and stored. As a result of such convenience heavy process machinery is able to function without interruption.

Also involved in this comfortable scenario, however, is an important ecological hazard. The most resistant component of wood, lignin, remains at the end of the chemical processing as an unwanted byproduct. Nearly all lignin which is extracted from wood in the course of paper manufacture is actually flushed into rivers, streams and lakes, where it can produce extremely severe pollution problems.

The precise lignin content of trees does vary from species to species. Fortunately it is lower in softwood trees, where it represents 15-20 percent of total mass, while in hardwoods the proportion rises to 30 percent or more.

It is the cellulose of the tree which the paper manufacturer requires. In all, more than 60 percent of the original tree goes in waste to produce the 40 percent of cellulose finally obtained.

Yet we must not imagine that paper is the only industrial product made from wood pulp. Other very considerable industries also depend upon cellulose. Many synthetic polymers, for example, are made wholly or partly from cellulose. Rayon (not just a clothing textile, but essential in the manufacture of tires) is dissolved and reconstructed cellulose. Treated and mixed with nitric acid and camphor, cellulose becomes celluloid. Other derivatives include explosives, lacquer, modern fireproof cinefilm, and many textiles and molding materials. Cellulose triacetate becomes "Tricel." Cellophane is another equally famous derivative. Water-soluble carboxymethyl cellulose is the thickening material commonly added to canned creams, mayonnaise and the like. Finally, wood itself, in the form of sawdust, is an important "filler" used on a vast scale in the plastics-from-petrochemicals industry.

This technology of making plastics and other synthetics from wood-based feedstock or "silvichemicals" is well understood and relatively simple. However, for the moment, coal-, oil- and gas-based chemicals have largely supplanted the use of silvichemicals. But whenever as little as a penny is added to the price of oil, silvichemicals become significantly more economic. And when we understand that fossil fuels will be substantially depleted in the next few decades, wood and wood products emerge again as natural front-runners. Not only synthetic textiles, but fuel for our cars and synthetic food may shortly be derived mainly from wood.

We can actually go further and say that there is scarcely one aspect of modern life that could not be catered for by forest products. Fortunately for us, and unlike coal, oil and gas, trees are a renewable resource. Or are they?

The stark fact of the matter at present is that while the demand for wood steadily rises, the creation and maintenance of reserves of living trees is becoming an increasingly unpopular pursuit. This is true not only of the Third World — where perhaps some kind of short-term economic excuse exists — but, tragically, of the developed countries as well.

The highly mechanized softwood forestry concerns of Scandinavia, the USSR and North America nowadays fell trees of an age which would have been rejected as unmarketable a mere twenty or thirty years ago. This development is not simply the result of improved technologies for processing timber, but a firm sign that increasing overheads and cost inflation are forcing producers to reap what hurried profits they can. Producers find themselves forced to pay less and less attention to the long-term quality and quantity of their stock-in-trade.

Often-cited reafforestation programs in countries like the United Kingdom offer no reassurance. First, such countries can accommodate only a tiny population of useful trees under present approaches. Second, there are also strong economic drawbacks. For example, these current programs are brought into being only through heavy direct and indirect state subsidies. The doubtful economics of the projects are often justified by appeal to — it must be said — rather nebulous amenity arguments.

The simple fact is that it is becoming steadily cheaper for industrialized countries to import their timber and wood pulp requirements from tropical Third World countries than it is for them to maintain forestry interests of their own.

The position just described is an incipient movement rather than a fully established trend. But already it is taking more permanent and concrete form, with the construction in several Third World countries of very large pulp-making plants funded by foreign capital. In any case, virtually all the craft needs for hardwood in western countries are already met by tropical imports.

Why exactly, we have to ask, is temperate forestry in such dire straits? What makes trees such an apparently burdensome crop to grow?

The time taken by trees to reach marketable size is the most important factor — especially during periods of economic inflation. Twenty to forty years of growth is required for most softwoods, while hardwoods require much longer still. All the costs of at least twenty years of planting and tending, plus the final marketing operation, must be met by the sale price. The market value of a timber crop must rise by at least 10 or 15 percent a year to justify the original investments and to add up to a going concern.

The conventional solution to these problems has been mechanized planting and harvesting on a grand scale, supervised by a small labor force. Once set in motion — that is, once the first twenty-year cycle is completed — such an enterprise generates cash flow sufficient to maintain the cycle of harvesting and planting indefinitely.

However, overhead costs — fuel, machinery, pesticides, transport — are harder on the larger concerns than on the small, especially if inflation causes running costs to rise at more than 10 percent a year. Or when taxation, especially if geared to short-term agricultural production systems as in Great Britain, is rigidly levied on invisible profits.

Not for these reasons alone, however, does forestry remain uneconomic, despite the ever-rising value of timber. The malaise is more subtle and deep-seated. It rests on the amounts of money and expertise spent on timber research and development.

Low levels of research and development investment are symptomatic of industries which consume natural resources. Their finances are based to a great extent on retrieving something for nothing, and on retrieving a something which is essentially finite.

But timber is, in principle at any rate, a resource which need never come to an end.

Once again these statements do not exhaust the complexities of our attitude to tree research and development. For even if timber were a nonrenewable resource, less is spent on research and development in the field than is spent on research and development into genuinely finite resources. Timber is still the poor cousin even among exploiting industries. Other industries which demonstrably have less of a future actually spend far more both on ensuring supplies of their basic materials and on developing more efficient uses and applications of the finished products.

Many undertakings with an obvious future (of which forestry, as we argue, could be one), such as the computer, instrument and drugs industries, spend between 5 percent and 8 percent of sales revenue on research and development. The aerospace, automotive and telecommunications industries spend less — 2-3 percent. Metals, mining and textiles all spend about 1 percent. Construction industries spend slightly less — but only one-fifteenth of this amount goes into timber research and development.

The last statement is even more surprising when we appreciate that most industries earn around a 30 percent rate of return on money spent in this way — often more than twice their return on capital investment. So there is, therefore, nothing inherently uneconomic in the idea of pursuing, say, a vastly improved pest-management technology for timber, or a full-scale production technology for silvichemicals. The block is psychological rather than economic.

A further complicating factor in our examination is the considerable psychological confusion over what kind of concern forestry actually is. Since forestry cannot be automatically bracketed with agriculture, is it then an industry? Once again, not automatically, since industries are not as a rule directly involved in the gathering of a harvest. So is it then rather a "something else," an uneasy hybrid about which we therefore have no ready-made ideas to help it earn its living or evolve its technology?

There are, as we have seen, strong environmental arguments — both local and global — in favor of conserving the world's tree populations, that is, of increasing those populations enormously in order that in thirty to fifty years' time there may be as much as half as many trees at large as there are today, instead of less than a quarter as many. Yet, the ultimate decision about whether or not trees can be conserved in quantity rests on financial arguments. Committed environmentalists may find themselves out of sympathy with such arguments, but they must consider them urgently above all else. They must identify those economic arguments which best serve the interests of the environment. In this way they will be doing everything possible to impose their logic on future events.

It now seems likely that it is too late to save trees in the highly concentrated form of natural tropical forests from near elimination. It may still be possible to set aside large tropical forest reserves as gene banks, leisure resorts and as habitats for endangered species of animals. Here a certain amount can be achieved by political pressure. We can also hope to persuade authorities that, in areas where tropical forests have already been cleared, commercial forests should be planted on at least some of this land, and hopefully on most of it.

Whatever happens, however, we must accept that the gross contribution the tropical forests have made to the global environment (as oxygen sources and carbon dioxide sinks, for example) can no longer be counted upon under present practices, certainly not beyond the end of the present century. It is time therefore that the onus of preserving the environment by conserving trees was shared by the developed world where, in a sense, the problem of deforestation began.

With the right strategy, we can afford in economic terms to correct the mess we have made even of our temperate native forests. We must also, as a matter of urgency, make some amends for the destruction of tropical forests and the more serious effects of their absence on the global environment.

To make forestry profitable in developed countries is an extremely tall order under current economic conditions. As we saw, the developed countries are turning increasingly to Third World sources not simply for timber supplies but for wood pulp. Afforestation programs in developed countries, if they begin from scratch, need discounts of the order of 3-5 percent in the form of tax relief or subsidies to top up the rate of return on investment to a reasonable level. The nonmarket benefits usually vaunted as good reasons for such special treatment amount to the amenity value of forests as recreation centers, tourist attractions, wildlife reserves, watershed shields, erosion defenses and the like. Yet, viewed dispassionately, the logic of the environmental case for commercial forestry is often suspect in a western context. The majority of Europe's and North America's commercial forests are not crucial to watershed or soil management. They are also probably not worth considering as gene banks, except in a very restricted sense. Nor do they actually have enormous value as leisure centers. Vacationers and tourists do not exactly flock to the dense shade of the vast conifer plantations which form the most economic forestland. Actually the edges and clearings of the commercial forest are far more important amenities in this respect than the forest itself.

Competition with agriculture for fertile land drives forestry up onto hillsides and into wastelands where the soil is marginal or useless for farming. In many cases, therefore, conifers are the only trees hardy enough to grow well in such situations. Though they are not the most appealing of trees, public opinion does tend to overdo its distaste for conifers. These can make forests which are as attractive in their way as mixed or wholly deciduous woodlands — but in practice they seldom do. The mix of trees chosen for commercial plantations is nowadays heavily determined by financial necessity — how long it takes this tree or that to reach maturity, how easily it can be machine-handled, and above all, how resistant it is to disease. A natu-

ral distribution of trees in keeping with the landscape, local conditions and public tastes cannot be easily achieved under such
constraints.

Because commercial conifer forest ecosystems are so homogeneous, the wildlife they can accommodate also has little diversity.
The forest floor itself has a low level of productivity on account of its
perennial dense blanket of slow-rotting needle leaves, so that the
absolute quantity of wildlife in conifer forests is also small.

Finally, fire and disease hazards are much more of a threat in continuous dense stands of conifers than, on the other hand, in small
mixed woodlands irregularly interspersed with clearings.

There are long-term economic drawbacks to afforestation which
have already been mentioned in general terms. The continued use,
for the time being, of mineral and fossil fuel reserves will inevitably
continue to undermine the finances of mechanized forestry.

So, paradoxically, a point is in fact reached where monoculture on
an industrial scale can show no economic advantage over old-fashioned, labor-intensive silviculture in small-scale mixed units. There
remain, as always, plenty of good reasons to plant trees — but, we
now ask, do they have to be planted in the form of forests?

Colin Rice(24) shares the view of a growing number of foresters
and agricultural economists who believe that a small-scale mode of
timber production could be perfectly profitable even today if it were
practiced widely enough. He maintains that:

> The cost to the future of exhausted mineral resources may be such that
> the very resource-intensive industries of a modern economy, e.g.,
> petrochemicals and construction, have negative overall value.
>
> . . . . If labour would otherwise be idle or engaged in material-extrav-
> agant production: if great fuel- and metal-consumption by forest
> machinery can be avoided, then it is worth producing wood, even at
> fairly low prices.

A reasonable extension of Rice's argument is the advocacy of agro-
silviculture, a system western technologists often preach to developing countries but apparently take no pains to encourage at home.

The total argument for placing silviculture firmly in the camp of
agriculture rather than industry offers not just good solutions to the
profit-making puzzle, but also a wide range of fringe benefits to the
farmer and public at large.

Woodlands planted in long parallel or contoured belts between
agricultural crops need not impair the efficiency of farm machinery,

as existing hedgerows and woods are liable to. If aligned at the right angle to prevailing winds such miniforests (or super-hedgerows) could greatly improve crop yields by reducing evaporation losses from the soil. They would probably also reduce crop losses occasioned by pests and diseases. For the rancher, the dairy farmer or mixed farmer, they could, besides improving the quality and quantity of pasture grass by their shelterbelt effect, become alternative sources of cattle fodder or even the basis for an integrated system of three-dimensional farming (see p.147). On irrigated farmlands, or those requiring constant drainage, the contoured belt reduces soil runoff and minimizes the cost of water-course maintenance.

These millions of populations of thousands of trees would also be easier to protect from disease, fire and other sources of destruction than are the huge forests of many thousands of acres. The mix of trees planted in the contoured belt could probably be varied considerably without loss of productivity.

As leisure amenities these belts could readily be styled to provide havens for a far greater number of leisure seekers than could ever be the case with conventional full-scale forests.

Other advantages follow. Wildlife refuges could be maintained as part of agro-silvicultural systems with little need for expensive special facilities or policing.

Then, because the leaf surface area exposed to the sun in a small wood is much greater than that of an equivalent section of deep forest (for, as stated before, in deep forests only the top of the tree receives adequate sunlight), agro-silvicultural plots would act as far more efficient oxygen sources and carbon dioxide sinks than today's conventional forests. Western agriculture could then begin to make a significant contribution to the global budget of atmospheric gases.

And to all these substantial benefits would be added local profits from wood production, not to mention national benefits in the form of saving on timber imports.

Of course, some of the above are rather simplistic statements which skate over detailed economic arguments, and depend on assumptions (albeit widely held assumptions) about unknown quantities — such as the limits of fossil fuel and mineral reserves. Even so, the case for agro-silviculture is no more beset by logical flaws than the case for expanding imports of tropical forest logs, or for continuing to segregate home-grown trees in vast, barely profitable monocultures.

The feasibility of agro-silviculture or any other novel mode of timber production would certainly be easier to believe in if this were linked with changes in the way timber is consumed. More specifically, if waste of wood could be avoided, there would be less need to impose current industrial standards of economic growth on the production of timber to meet our essential needs.

We all waste wood. Most of all, we waste it in the form of paper. Timber, however, as we saw, is not essential to the production of paper. Any vegetable material can in theory be used instead. It is even technically possible to manufacture paper from processed sewage, though the actual quality is not all that impressive. Yet wood pulp remains at present the most economic raw material for paper production, for the kinds of reasons we have already discussed. There is, however, still no good reason why it should go to waste in that particular form to the extent which it does today. Paper can be economically repulped and recycled several times over — but instead billions of tons of fresh wastepaper are daily thrown away to rot or burn on garbage heaps. There is not so very much difference between this practice and the slash-and-burn cultivation that is deforesting the tropics — except that in the latter case at least, some palpable benefits, however temporary, arise in the form of food production. All too often, however, the printed page fails to support the merest gleaning of hard information and food for thought.

Since literacy became a universal aim of political reform, the reading habit has become an intimate part of the lives of half the world's adults. In Europe and America the average consumption of words is probably around a million per adult per year. Aldous Huxley(11) pointed out that:

> To a considerable extent reading has become, for almost all of us, an addiction, like cigarette smoking. We read, most of the time, not because we wish to instruct ourselves, not because we long to have our feelings touched or our imaginations fired, but because reading is one of our bad habits, because we suffer when we have time to spare and no printed matter with which to plug the void. Deprived of their newspapers or a novel, reading-addicts will fall back on cookery books, on the literature that is wrapped round bottles of patent medicine, on those instructions for keeping the contents crisp which are printed on the outside of boxes of breakfast cereals. On anything.

Printed matter which aims simply to give or retrieve information — tax forms, textbooks, banknotes, scientific papers and so on — accounts only for a small fraction of the printed paper that is in circulation at any one time. Regardless of the advent of mass telecommunication, the great bulk of printed matter does not serve any hard practical purpose, but is either totally trivial or is given over to what Huxley would have called commercial, ethical or political propaganda. None of these, in Huxley's view, could ever have a mechanical efficiency of more than 1 percent. His estimate was intended ironically, but it is probably not very far from the sober truth.

This book is a piece of ethical propaganda and is, as such, no exception to Huxley's rule. But because it is printed on recycled paper it is automatically more efficient in its use of raw material than a book which is not. Moreover, since the cellulose fibers this page contains have now borne more than one message, the chance that the compound total of the subject matter carried might serve a useful purpose is at least twice as good, or less than half as bad, as it would otherwise be. By a "useful purpose" is meant a purpose that might compare in usefulness with the function of the roofbeams of a house — which is what many trees currently harvested for pulp would have become had they been left to mature.

As consumers of paper we may find it impossible to break the compulsive habit of reading rubbish (as opposed to useful reading), but we can at least help to retrieve wastepaper. We can accept the invitation of the Friends of the Earth and other prominent environmentalist groups to organize collection points for wastepaper, and press for official action to regularize and popularize recycling facilities. Not least to encourage paper companies to take recycling seriously. Therefore, of course, publishers and large business or state service organizations also have a part to play. They should consider recycled paper as a first choice for all print jobs, wherever this is feasible.

Timber and pulp producers can help boost the efficiency of forest exploitation by continuing to find ways in which the waste products — waste wood, foliage, bark and lignin — can be put to economic use. The development of new types of wood product (e.g., fiberboard, plywood) has already greatly enlarged definitions of the kinds and parts of trees that can be productively used. Such developments have, however, their drawbacks also, for they have helped to make forest-product industries still more capital intensive and machinery dependent. And to some extent forestalled improvements in the fundamental quality and variety of commercial forests, since they can make use of poorer and still juvenile trees.

The exploitation of wood as a building material, or even as a basis for paper or textile manufacture, need not in itself give the slightest

cause for ethical concern if it truly gave tit-for-tat in the form of recip-
rocal renewal and replanting programs. The evidence, however, is
very much to the contrary.

Managed forestry enterprises in the developed world almost
invariably give back to the environment far less than they take out at
harvest, no matter how numerically correct or generous their rate of
replanting may be. What little variety still resides in native forests is
rapidly becoming attenuated both by disease losses and by the planting practices designed to counter these. The selection for planting of
kinds of trees which are easy to protect from disease or which are currently disease free may have short-term advantages, but in the long
run the tendency toward monoculture must carry grave risks. By putting all their eggs in one basket, timber producers court disaster in
the form of unforeseen disease epidemics. The natural selection of
inherently resistant trees is too slow to serve the profit-making imperative in big forestry, so producers often fall back on an ever-dwind-
ling number of exotic varieties of plantation stock to keep
production up to par.

At the same time, the age at which trees are harvested is kept so
low, to minimize overheads, that the role of these forests as components of the environment is nothing like as great as it ought to be.

At least, however, trees are planted in big forests. Completely new
forests are also planted from time to time, though we have seen a
sharp decline in afforestation especially during the past few years. In
Britain, for example, the amount of land converted to forest in 1977
was less than 45,000 acres — some 60 percent less than the 1975
figures. In the United States, only 740 million acres were under forest
in 1977, as opposed to 754 million in 1970.

Most net losses of tree cover can actually be attributed not to big
timber producers but to the private tree owner and land developer.
These, by clearing woodlands and hedgerows (or failing to replace
them when they die), have placed in jeopardy the future of temperate
hardwood trees as a life form. Tropical hardwood imports, besides
aggravating the plight of tropical forest ecosystems, seriously mar
the economics of home-growing native hardwood trees, like the
beech, oak and elm. Epidemics of disease like DED, oak wilt, chestnut
blight and beech bark disease then greatly amplify these effects.
Where replacement of depleted temperate hardwood stocks is still
practiced, it increasingly involves replanting with low disease-risk
trees, like limes, southern beeches and sycamores — a depressing
imitation of the gene-extravagant retreat that is taking place in com-

mercial softwood forests.

Yet with different management practices, including the aforementioned change of emphasis toward agro-silviculture, both soft and hardwood forestry could begin to address disease problems in a more confident fashion. Decentralizing and demechanizing the production of all kinds of timber would, as emphasized, create opportunities to make a concerted stand against particular diseases.

Because it is already disposed in scattered, small populations, the elm and its disease problems constitute a test case for agro-silviculture. It has to be demonstrated that trees like the elm can be protected from disease in small woodlands, avenues and hedgerows, otherwise businessmen-farmers, however interested they may be in the potential of conserving trees, will understandably regard the risk of epidemics as a prohibitive unknown factor.

The possibility of overcoming DED is far from remote, as I hope to show in the next chapter. The disease is already being successfully controlled in isolated parts of Europe and America by cheap and straightforward methods. It is, I personally believe, only because DED-control strategies have not been promoted at national and international levels that the problem seems currently so intractable.

If large populations of elms cannot be saved either for love or money, we must not be surprised to see other trees sharing the elm's fate in one way or another. It will, in a very real sense, be our own fault if this happens. To avoid this disaster we may have to take leave of the mercantile approach which at the moment governs official policy toward tree conservation. But that is our prerogative.

There is a case for preserving the elm and there exist techniques which will enable us to do so. We must press that case and apply those techniques in a concerted fashion if we do not wish to look around us and find nothing to rest our eyes on but the straight line of the horizon, like the line on a cardiograph which shows that the heart of a living thing has failed.

10 A Future for Elms ...and for All Trees

"Thou'st sheltered hypocrites in many a shower
That when in power would never shelter thee."

JOHN CLARE
"The Fallen Elm"

GERALD WILKINSON'S BOOK *Epitaph for the Elm*(38) was published in 1978 From one point of view any book is good which draws attention to trees and, in this case, to the dangers which trees currently face. From almost all other points of view this book will not do.

For one, it accepts that the elm is finished — almost rejoicing in that fact, in the way that some individuals, at one level, savor the funerals even of those they have loved. The elm is dead, says the book. Let us, however, not grieve too long about that, but instead replace the elm with other trees — oaks, beeches, chestnuts, poplars.

Wilkinson entirely overlooks the fact that these other trees may also shortly suffer — *and in some cases are already suffering* — a similar fate. The chestnut is a particularly unhappy inclusion in the list, in view of the virtual disappearance of its American cousin.

Epitaph for the Elm is simply an admission of defeat and a reproach to every one of us. The more so because it is almost true.

Almost true, however, is not the same as true. We can, if we will, and without asking for miracles, make sure that it never becomes true. It is not a case for divine intervention, it is a case for human intervention. *The elm is for the saving.* All that we have to do is to make the necessary effort.

There are several theoretically possible ways of tackling Dutch elm disease. In each case we can and must as always substitute for "elm" the name of any tree under wide-scale attack from disease.

In summary, the three most likely approaches are as follows: (1)

We can remove diseased elms and replace them with new young elms; we would then however also have to take steps to see that these did not fall victim to DED in their turn; (2) We can treat diseased elms, so that at least a majority of mature trees are saved; (3) We can remove diseased elms and replace them with other varieties of trees, so far not the victims of epidemic disease.

There is actually a fourth approach; it is simply to remove the diseased trees and replace them with nothing.

The basic question is: should elms be replaced by other trees? A more practical one is: *can* elms be replaced by other trees? In both cases there is always the further implied question: how much is it going to cost?

Since the 1920s and 1930s, when the seriousness of the DED problem first became common knowledge, the creation of new landscapes and townscapes in place of those formerly dominated by elms has frequently been argued as the only businesslike alternative to (allegedly) costly and fallible DED-control and elm-replanting programs.

Lime trees, sycamores, southern beeches and various other kinds of tree have been publicized as desirable elm substitutes. Amenity planners have devoted a good deal of time and money to devising schemes aimed at using these trees to plug the visible gaps left by the death of elms in town and country.

As far as urban settings are concerned, such schemes are, of course, already long beyond the planning stage. In the USA, where elm disease has always been popularly regarded as essentially an urban problem, plane and small-leaved lime trees already outnumber elms in towns and cities where the elm reigned altogether supreme only a few decades ago.

Public attitudes toward this quite remarkable change in the way eastern American towns look vary enormously. T.W. Jones, an eminent United States Department of Agriculture plant pathologist, told me that in cities where most of the elms have already succumbed to disease, local people seem to have little hope of, or interest in, saving the remaining trees. In cities where DTD has only recently appeared, or where control programs have succeeded in keeping losses low, however, there is nearly always widespread support for disease-suppression measures. Here we see very clearly how hope dies along with the trees.

To be precise, in towns where annual disease losses exceed 10-15 percent of the total elm population, DED-control programs tend to lose their effectiveness or their credibility.

In many towns where disease losses remain below 15 percent per year, however, it is often demonstrably cheaper to keep DED in check than to remove large numbers of sick trees *en masse*, replacing them with young trees (elms or others) which will not provide beneficial amounts of shade until ten or more years after they are planted. It turns out that where the annual loss of trees does not exceed 15 percent per annum, a control program based on the prompt removal of badly infected trees, together with the treatment and protection of part-diseased or disease-threatened trees, is usually the more cost-effective option (see Figure 11). Some American cities in the heart of DED-hit states have managed, thanks to such programs, to keep annual disease losses to 3 percent or less.

It must be emphasized that because there is no centralized fund or national legislative strategy for applied elm conservation in the USA (or anywhere else in the world for that matter) local authorities who are prepared in principle to plan and pay for local conservation programs must justify them at every stage to a small electorate, and a sometimes impatient opposition lobby.

The feasibility of DED-control programs anywhere, then, can depend as much on local political conditions as on cost-accountancy. But cost usually decides matters. The expense of skills and materials appropriate to control programs can differ giddily from state to state, even from town to neighboring town. In the absence of a comprehensive national policy toward elm conservation (as distinct from the *laissez-faire* approach and the limited advisory and research service the USDA promotes in America, or the UK Forestry Commission and its counterparts in Europe), integrated field programs for practical conservation are at present out of the question. These basically involve close cooperation between adjacent local authorities and call upon standard procedures to be available at standard cost.

Owing partly to the parochial nature of existing control schemes, profiteering is rife among small tree-surgery concerns. These may, without scruple, advise the felling of perfectly healthy or curable trees or the expensive treatment of incurable cases, depending on the slant of local market values. And even though the majority of experts involved in the protection of amenity elms are quite scrupulous when giving advice, taking their advice still depends on values liable to change without warning (such as the interpretation of the benefit half of cost-benefit findings) long before an adequate control

Figure 11

Comparison of annual average costs of two methods of countering Dutch elm disease

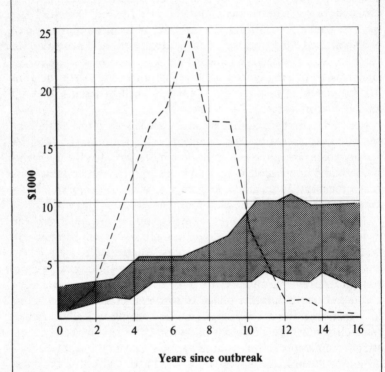

Years since outbreak

(a) No action beyond removal of dead trees as they become safety hazards (dotted line)
(b) Adopting active control measures (shaded area)
The model is a US municipality in a DED-outbreak area, in relation to a sample unit of 1000 trees.
Strategy (a) puts a disruptive strain on the annual budgeting of muncipalities which adopt it, though costs fall off as the elm population dwindles to less than 15 percent of original strength after about twelve years.
Strategy (b) enables costs to be spread over many years and offers hopes of retaining a substantial elm stock at modest expense. The cost of control strategies varies according to precise local circumstances.

program can be established and carried through.

If DED were exclusively an urban or amenity problem, this state of affairs would be merely one among many depressing aspects of the perpetual loss of character we see in the urban scene. Urban elms do not, after all, constitute natural populations. They exist in towns because each one of them was put there to provide certain benefits to town-dwelling humans: shade, shelter, ornament and maybe a few hidden extras such as protection from noise, pollution and airborne dirt. These are important boons but, in fact, there are many other kinds of tree which can provide them just as well as, if not better than, the elm.

Elms, including urban elms, also have an objective face value, usually computed on the basis of their potential real-estate or timber value, and this is sometimes cited in an attempt to tip the balance of argument in favor of urban DED-control programs, where a straight cost-benefit analysis does not add up to a good case.

The four hundred thousand or so elms known to die in the USA each year through DED were claimed, in 1976, to represent an estimated annual financial loss of one hundred million dollars. This figure is considerably, though not grossly, higher than the sum spent on local elm conservation programs during 1976. But nearly all the conserving was done in cities, towns and suburbs where it can have made precious little difference to actual elm timber or real-estate markets. Insofar as they can be grouped into an identifiable whole, elm timber markets obtain most of their supply out of town, where the bulk of elm stocks are always to be found. And the extra insurance premium placed on houses with trees growing near them, which are liable to undermine foundations (the elm is notorious in this respect), can outweigh, in practice, any difference to property prices which attractive elm avenues may make in an urban neighborhood.

The prosperous towns and cities, where numerous elm avenues and parks are still actively ecouraged to survive, provide at present, for what it is worth, the nearest thing to hope for the ultimate conservation of some western elm species. They are (in a manner of speaking) the world's only sizable elm reserves, where DED-control programs can still be expected to attract a flow of publicity and practical resources, if luck and local support remain on their side. Yet is this situation, where such money as is available is spent on nonessential urban elms, justifiable?

It is hardly surprising that public initiatives to conserve elms should be strongest where public wealth and opinion are at their most concentrated. It is understandable that significant numbers of people will show they care about the consequences of a catastrophe they can see from their windows. Yet is it realistic to speak of the loss of elms from municipal settings as a catastrophe?

The plight of elms in natural and agricultural ecosystems should, for many reasons, give considerably more cause for public concern. Through this chapter I shall pursue the point that the fittest arena for a battle to save elms is not in the street, but in the field. And I shall describe flaws in existing elm conservation programs which support a case for a more pragmatic and broadminded approach to the task of helping elms hold on to their rightful place in the environment.

The demand for trees in towns, though it springs from somewhat ill-defined motives, is reasonably constant, reasonably well-entrenched in the ethics and aesthetics of urban planning. No town-plan blueprint is nowadays complete without an incipient rash of tree symbols. Many of these trees mysteriously perish on their way from drawing board to boulevard, but the garden city ideal retains, nonetheless, a firm hold on successive generations of town planners and town users. It is rare, for example, to find a municipality which does not replace most dead shade trees as a matter of course, even though the expense may be considerable.

Outside municipal limits, however, the reverse ethic applies. Here we find a general need to constantly increase the area, as well as the yield-per-acre, of agricultural fields. Trees are a barrier to this form of progress. They block the free movement of machinery, the acreage they occupy could grow corn, soy beans, and so on.

There is little, and decreasing, hope of finding in farming areas today those kinds of common land or no-man's land where elms characteristically thrive. There are no boundary lots, small woods and waste plots near villages and farmsteads. The past few centuries have seen a rapid decrease in the availability of such habitats throughout the northern hemisphere, and a consequently sharp decline in the numbers of the trees which typically exploit such marginal sites. The added disappearance of agricultural land itself beneath urban and industrial sprawl (an estimated fifty to sixty thousand acres annually in the UK, much more in America) applies yet another turn to the screw.

There is, in other words, a universal bias against the retention or replacement of farmland trees whether they are healthy or sick, and whether they are elms or pines or oaks. An outbreak of DED can often provide the farmer with a convenient opportunity and a plausible reason to remove woods and hedgerows apparently forever. Even in situations where trees still serve an undoubtedly useful purpose (such as forming part of an important boundary) this bias still operates after a fashion. For instance, the cost of erecting and maintaining wire fences is no lower, in most places, than the running-cost of a well-kept existing hedge-and-ditch system. But if it frequently becomes necessary for the farmer to lay out funds on the treatment, felling or replacement of a stock of epidemic-prone trees, the balance can easily tip from them in favor of a capital layout on hedge-removal and the installation of artificial fencing.

Any low-key local initiative to conserve rural elms can thus be countermanded by all-pervading economic imperatives to dispense with trees in general. This is one large pinch of salt that has to be taken with adventurous-sounding menus for novel landscapes framed by new kinds of tree intended to take the place of the stricken elm.

A relevant consideration, which such schemes usually also leave out of account, is the unlikelihood that tree species so far recommended as elm substitutes can fairly match the specifications of healthy elms as farmland trees. Those specifications include few attributes which a professional horticulturalist might defend, but plenty which, in an agricultural context, make elms better stayers than many, perhaps most, other trees.

There are three practical ways in which elms can have the advantage over other types of tree.

In the first place, it is no coincidence that the types of elm which show a stubborn tendency to increase their numbers by vegetative means, such as suckering or layering (i.e., sprouting from the roots or grounded foliage of a parent tree), are particularly common in farmland. It is an example of how natural selection in agro-ecosystems really amounts to agricultural selection. In an English elm hedgerow, for example, there is a constant burgeoning of aspiring suckers that contribute mightily to the thickness and strength of the row as a whole and line up to take the place of their mature parents when the parents die. Those shoots which grow out of line with the hedge are soon eradicated by plowing or grazing, and though their shallow roots are occasionally liable to damage machinery, they do not present really serious problems to the farmer. Certain other trees, however, which eschew suckering and instead cast fertile seeds far and wide, give rise to deep-rooted seedlings which can cause weed problems (a weed being defined as a plant in the wrong place) over a large area. This holds especially true if their germination predates or postdates plowing operations. In the one case they will be plowed deeper and produce yet deeper-rooted weeds. In the other, they spring up in the midst of a growing crop from which they cannot be removed without great inconvenience.

Again, to make nonsuckering trees grow in a disciplined, dense row can entail added labor, nurture and expense on the farmer's part — sufficient perhaps to jeopardize the desirability of having a hedge at all. Several tree species (such as the sycamore) which have been publicized as elm substitutes pose these kinds of problem. Giving elm suckers an even break has the great advantage, from a farmer's point of view, of saving labor. In their healthy state, farmland elms are, at the very least, easy to tolerate.

Secondly, the drawbacks which can be listed to the discredit of the elm's heirs presumptive, though seemingly minor when viewed singly, together add up to quite solid disqualifications. For instance, the leaves of alternative trees, if they are large and slow to rot, can boost the acidity of the surrounding soil and oblige the farmer to spread extra lime around field margins. This problem assumes greater importance in regions where soils are already low-alkaline.

Lastly, nonsuckering trees may run into a snag quite the opposite of the weed problem mentioned earlier in areas where insecticide sprays are intensively used. Their seeds may fail to germinate because of the scarcity of pollinating insects in the neighborhood. Farmers and estate owners may in this case have to invest in nursery-grown trees if they wish to maintain a mix of generations in their hedge and ensure its continuity.

In purely practical terms, then, elms often have persistent advantages over other kinds of tree as a component of agricultural systems, even if some of the uses for which their ancestors were actively cultivated are partly or wholly defunct. Elms are often simply deemed less trouble to have around than to exclude. This statement of course ceases to be true when disease prevents elms from looking after themselves. They then become a serious liability and nobody can really blame the farmer who wants to be rid of them — except on this ground: that their fate is not a matter to be considered only in the light of obvious agricultural efficiency or of the composition of landscapes. There is far more to the presence of elms in the countryside than meets the eye. The long-term effects of the removal of elms on the ecological balance of long-cultivated areas has to be taken into account — though here our knowledge is lamentably imprecise.

These we might call the hidden benefits of elms.

Lists can be made of organisms thought to show specific preference for elm habitats or nesting sites and known to be predators or parasites of common crop pests or weeds. I mentioned in an earlier chapter (p.73) that the decline in numbers and the redistribution of populations of certain birds in the UK may be one ecological side effect of DED epidemics. Such consequences may have a considerable economic impact on farming in areas where elms were formerly abundant. Still more sensitive studies of ecological relationships between elms and other organisms (including types of crop and farmland animals) would probably show many invisible links that are very important indeed. Matters could hardly be otherwise in view of the length of time the elm has played a dominant part in agricultural vegetations — some ten thousand years or so.

Equally, however, there exist some pests (e.g., mite pests of orchard trees) which may in the course of their life cycle be hosted by elm trees. The precise relation of beneficial to harmful effects of the elm is not known at present. But the very abundance of elms in particular regions through century on century of cultivation speaks rather of harmony and of positive advantages than the contrary.

For example, the early cultivators of grapes had no doubt that elms were a good influence on their crop. St. Jerome wrote in the fourth century A.D. of the "great affinity between elm and the vine," and the practice of actually training vines on elm trees only ceased to be normal in some regions of Europe about a century and a half ago, when viniculture became too intensive to wait on slow-growing, three-dimensional vineyards.

Many other examples of such affinity between kinds of plants, going well beyond mere coincidence, are well known. Plum trees and blackcurrant bushes, for instance, always grow and yield better when planted together than when grown apart. The exact nature of such relationships is for the moment only guessed at. It is probably based on the ability of the twin crop plant to produce a surplus of nutrients useful to the other, or to harbor organisms which feed on the other's pests and weeds.

Yet if elms were known for certain to exert an extremely beneficial influence on this or that crop, such knowledge might not even then by itself amount to a sufficient case for preserving and rehabilitating elms in the countryside. Still, it is one arrow in the total armory of argument. It does, of course, not do to talk of the environmental role of elms as if this were in any case a simple uniform phenomenon. Different species and samples of elms undoubtedly play different parts in different ecosystems. It is hard enough to make an intensive ecological study of any one single variety and its dependent organisms, let alone establish a body of knowledge that can explain the ecology of entire elm populations, no matter where.

Concern for beauty and loyalty to heritage are the commonest motives shared by people and organizations involved in existing practical elm conservation efforts. Although economic and ecological arguments are often advanced in the prospectuses or press statements that may precede or accompany such rescue attempts, it is pointless to deny that the main reasons people try to preserve elms are, for want of a better word, sentimental. It would be foolish to belittle them as such. For such reasons make just as much sense as other sets of values which, whether or not we are aware of it, govern our day-to-day decision making, getting and spending. Moreover, sentiment in this context can paraphrase a certain kind of environmental horse sense which everyone possesses to some degree.

As a basis for decision making at a national level, however, sentiment and horse sense have their drawbacks. They count far more as a pressure to maintain an existing local status quo than as a brief for entirely new national policies needing the approval of parliaments and senates.

So it comes about that individuals who are prepared to go to great lengths to ensure that their children and grandchildren may enjoy a surrounding of elm trees have little choice but to concentrate their efforts upon creating — in effect — elm reserves in their own neighborhood.

How should we define the ideal elm reserve? Would it be so different from those which already exist?

First, it should be sited where environmental conditions are not friendlier toward DED than toward elms. It is likely, for example, that elms growing in a place where groundwater levels are low or prone to unpredictable fluctuations will be less generally fit (and therefore less likely to survive or resist disease) than if they grow in a spot where the water supply is good and steady. Quality of soil is also vital. Urban elms, no matter how frequently and carefully they are tended, are underprivileged in both these respects. It is true they are tough enough to grow well in the meager, hard-packed and infertile soil of city streets, sandwiched between the power mains, drains and sidewalks. They even somehow manage to intercept rain water before it disappears into the sewers. No doubt about it, they are hardened townees. But it is unreasonable to ask them to make a stand against a disease epidemic under conditions which already stretch their capacity for survival to the limit.

The ideal elm reserve should also be sited where the trees are to some extent isolated from other areas of distribution both of elms and DED itself. In other words, it should comprise a pocket of elm trees in a larger area where elms are uncommon or where geographical barriers help prevent the spread of bark beetles.

A classic strategy of limiting epidemic tree diseases is to create a *cordon sanitaire* around threatened populations, a buffer zone divested of every actual and potential source of infestation. This technique is often employed in large forestry plantations, where it can be planned and policed efficiently. It is less easy, of course, to bring this strategy to bear as precisely in towns or upon scattered populations of rural elms. But some form of sanitation clearing is absolutely vital for workable DED-control programs. In practice, sanitation usually amounts to a version of the *cordon sanitaire*.

The importance of the *cordon sanitaire* cannot be overemphasized. We see its absolute necessity in cases where one local authority makes inadequate efforts in an area adjacent to another authority's sound and perhaps expensive program. The former area serves as a breeding ground for disease-carrying beetles, so that all the work of the conscientious neighbor goes for nothing.

A good example of a successful sanitation program is found in the rural area of East Sussex in southeast England and its adjoining municipality of Brighton. Local authorities in these two districts, one town and one country, began to implement a rigorous sanitation policy. This they coupled with a well-publicized early warning system that relies on the help of local people. The scheme was begun in the late sixties. Elm losses through DED have since been reduced in both areas to only 2-3 percent a year. In neighboring West Sussex by comparison (where the budget for DED suppression is approximately one-tenth of that in East Sussex) losses currently run at 40 percent a year, and already more than 80 percent of the preepidemic elm population has disappeared.

This striking contrast between these adjacent areas would be hard to maintain if the protected area of East Sussex was completely bordered by similar DED-breeding pools. But the boundary of East Sussex is mainly coastline. With the sea thus providing most of the *cordon sanitaire*, it continues to prove feasible to patrol the rest at public expense.

The apparently sterile limits of a town or city, incidentally, are not an effective geographical barrier to the spread of DED. The ability of elm-bark beetles to home in on a host tree can be phenomenal. I have seen solitary elm trees in London back streets several miles from the nearest source of infestation (and on a smoggy day) aswarm with newly arrived *S. multistriatus* adults. The perpetual, hurried circulation of materials around a densely populated town or city is also extremely favorable to the accidental spread of DED and its carriers.

A special disadvantage of tame elm populations is that individuals tend to be planted rather close together. DED is obviously more liable to spread quickly through densely packed ornamental stands and rows of elms than it will through scattered wild populations. We know, of course, that many countryside elm populations are anything but scattered. Indeed, a rural elm hedgerow is just as easily overpowered by DED as an urban elm avenue would be, particularly if the infection is root-borne. *En bloc*, however, even rural elms crammed together in hedgerows and inseparably linked via their roots tend to have a certain advantage over their urban counterparts, in that they are usually surrounded by a mixture of trees and plants other than elms. Thus bark beetles meet with more distractions in their search for hosts and are more likely to fall foul of the wider range of predators (especially birds) that one might, in any case, expect to find frequenting a mixed vegetation.

Town elms tend also to lack genetic diversity within their own ranks. Most of these are grown from commercial nursery stock, and while their numbers may well include cultivars specially bred for DTD resistance, the total number of different genes in circulation among them is probably smaller than among equivalent wild or semiwild populations. It is not unusual to see, even in a clone of wild elms which have suckered from a single ancestor, genetic mutations at work which apparently confer some degree of natural disease resistance. The sight of one or two healthy trees in a dead or badly diseased row testifies to the reality of this phenomenon.

Resistance-bred cultivars might, at first thought, be expected to have such useful mutations already plentifully etched into their heredity. But the fact remains that no universally resistant elm cultivar has yet been produced. The main problem is that the physiological basis of DED resistance is still not fully understood. Size of xylem vessels and the timing of tylosis (see Chapter 4) are thought to be among the most crucial factors. Elm breeders can certainly take vessel size into account when they select their breeding stock (though in practice their approach is often more empirical), but nobody has so far developed a simple technique for relating the rate of infection to that of vessel tylosis.

If the timing of tylosis turns out to be the most important difference between DED-resistant and DED-prone elms, resistance breeding programs may have to undergo a considerable change of tactic. For it is quite on the cards that environment is just as important as heredity where the tylosis reaction is concerned. In other words, the same species or variety of elm probably undergoes tylosis at different rates in different soils, climates and surroundings. When natural resistance crops up in wild elm populations, it can, at least, be reckoned to be the right answer to the form of DED prevalent in that particular locality.

The current policy of concentrating DED research funding on breeding programs in a few generously endowed research centers, while leaving the bulk of the natural elm population to fend for itself, is therefore open to very serious doubt. It is not just a case of putting all our eggs in one basket. There is no ultimate guarantee that the eggs are even good eggs. It is, in fact, by no means certain that any amount of research will eventually produce "super-elms" capable of resisting DED under any circumstances.

Yet even if such super-breeds were eventually developed, it is naive to imagine these would be planted out in numbers that would anything like meet current replacement needs. A saner alternative seems clear. If, using the same funds, existing populations and groups of wild elms were artificially maintained at present levels, these populations would function as natural laboratories geared to the production of DED-resistant — more especially, locally resistant — trees. Given artifical help through the worst of the crisis, existing trees can and will produce the mutations we require.

In addition to specific well-founded initiatives of this kind under present crisis conditions, there is every reason to suppose that we could reduce the crisis level itself. It is not seriously to be doubted that DED could be brought under control and eventually reduced to nonepidemic levels by an integrated national pest management approach.

It is the feasibility of a national control program we must now consider, in the context of the creation of large national elm reserves.

There is nothing basically new in the idea of such programs. The government of every country affected by DED has at one time or another been urged by conservation interests to adopt nationwide control strategies. But no government so far can be said to have responded at all adequately.

The main reasons for the failure to act are often more complex than might be expected. Cost is not necessarily the most intractable of them. Something much closer to a national program could be achieved in many countries by a straight redeployment of the public funds already spent on DED research and on elm conservation in national parks.

I have already stressed how difficult it is to press a reasoned case for national investment in uneconomic trees. Economic considerations similarly obstruct moves to conserve any form of wildlife in any industrialized country. It is, even so, easier to recommend and create legislation to protect small and rather obscure organisms — rare hawks and butterflies, for instance — from extinction or harassment. But huge and ubiquitous creatures like elms, or whales, are another matter. Their protection is bound to involve substantial trouble and the reasonableness of legislation for a dose of trouble has to be placed beyond doubt. If the issues or our proposals for meeting them are fuzzy, the political response is all too liable to fuzz in return.

A third major problem is the *amour propre* of established specialist interests. Token responsibility for finding a solution to DED epidemics has so far been entrusted to broadly based national forestry services or agricultural research agencies, bodies reluctant to wonder whether the problem might lie outside their scope and quick to believe that their expertise or integrity is being called into question, when nothing is actually further from the truth.

Despite these obstacles, I believe that there is still hope for a national or international approach to the task of securing a future for elms. As a basis for discussion, I shall outline a notional strategy founded on the creation of large reserves of elms in carefully selected areas of all DED-affected countries.

Comprehensive surveys and censuses of natural elm populations and their settings would first of all identify the areas best fitted as reserves. These would possess qualities (like those listed earlier) likely from the start to hamper the successful spread of disease. As far as I know, elm surveys have only been made to any broad extent in Britain and perhaps in Holland. Yet useful survey operations need not be prohibitively awkward or expensive to organize. High school biology departments, large conservation and amenity-protection societies, and other volunteer groups would, I feel sure, respond to well-publicized requests to cooperate. Together with scientists and foresters, they would gather raw facts about which elms grow where. Such groups would, I suggest, be briefed by, and report back to, a centralized organization charged with establishing statistical analyses of survey results. Those areas of optimum conditions for the creation of elm reserves would be identified. This same autonomous and purpose-designed body would then be responsible for the actual demarcation of a number of reserves, each incorporating at least fifty thousand elms and positioned if possible at optimal distances from each other across the national map.

The reserves would be artifically isolated, as necessary, by a *cordon sanitaire* up to ten miles wide, where all other elms, sick or healthy, had been removed in the interest of protecting the reserve from epidemic influx. Agreed amounts of cash compensation could be paid to landowners to whom this activity represented a genuine hardship. All transport of timber in and out of the areas would be subject to strict mandatory checks aimed at preventing the accidental distribution of infection.

Inside the reserves, rigorous sanitation initiatives would eradicate all existing evidence of DED. Reserved elms would be kept up to their original number by the replacement of each dead tree with young elms of a mixture of species and varieties known to have resistant qualities — and preferably with seedlings or suckers of apparently resistant stock growing in or near the reserve itself.

Experimentation with fungistatic injection and other techniques would also be encouraged under the supervision of trained wardens. Exact studies would proceed in parallel to determine the ecological importance and potential economic applications of elms. Each reserve would host its own study facilities, nurseries and hardware.

Leaving aside for the moment broad questions of feasibility, we can ask what advantages the strategy described might have over existing, randomly localized attempts to preserve elms.

The most obvious advantage, in my view, is this. If DED-resistant stocks of elms should ever be developed by one means or another, or if DED itself died out with or without prompting (a possibility which can never be completely ruled out), it would then be relatively easy for the elm to become significantly redistributed across its former natural haunts from these numerous, scattered vantage points in ordinary farmland and woodland — far easier than from a few city enclaves or experimental nurseries. Meanwhile, an accurate knowledge of the ecological and economic importance of elms would have been gained from studies made under genuine field conditions, untroubled by an immediate local threat from epidemics of D ID.

If, after some agreed experimental period, the reserve system did not appear to be making clear progress in the hoped-for directions, we could at least feel assured that we had done our utmost in our own and the elm's interest. We would certainly have established techniques and precedents with which to combat other tree disease epidemics before they attained the dimensions of the current D ID problem.

MAINTAINING THE NATIONAL RESERVES
So who could run a system of elm reserves and who would pay for it?

I would argue that DED is too universal and perplexing an issue to lay at the door of any one professional body or any one group of interests. It is impractical to expect farmers, or landowners, or research scientists, or timber traders, or conservation groups alone to take the rap for what is, after all, a national and international ecological crisis. I use that term advisedly, for a crisis in the biosphere does not have to pose an obviously direct threat to the human condition to be a crisis. Elms, like all organisms, inhabit a world of their own and respond to it in their own absolutely unique fashion. They have as much right to occupy space on the planet as we do — far more really, if their contribution to the well-being of their fellow creatures were measured and compared with ours.

We are very much to blame for having put the survival of elm trees at risk and our obligation to make amends is great. Yet just as no single human agent or profession can be impeached for having helped DED to become the success it is, so it is appropriate that the search for ways to reverse that success should be shared by all.

I am not now simply talking about sharing the cash cost of DED-control programs. I believe it to be just as, if not more, important that as many people as possible should actually participate in these programs to the best of their ability.

Members of postindustrial western societies are in any case constantly being made aware that there are limits to an exclusively technological approach to problem solving, even when the problem is apparently and essentially a technical one. It has been well said (by the physicist John Ziman) that the trouble with technologists is not that they provide wrong answers or that they cannot be trusted not to feather their own nest, but that they often fail to pose the right questions in the first place. The Concorde project is an apt example

of this conceptual flaw. All the designers, technicians and economists involved in creating the Concorde contrived very successfully
to answer the question: was it technically feasible to put a supersonic
passenger transport into service? But they did not consult potential
customers or the interests of people living near large airports. As a
result, British and French taxpayers are left with a piece of equipment which is (in economists' jargon) "negatively profitable," to the
tune of several million dollars a day.

In Ziman's view, the way to avoid such errors is to involve as many
different interests as possible in the planning and realizing procedures conventionally farmed out to technological pundits. We
should ". . . exploit the educational potential of working together"
— not automatically commit resources to technologically optimal
solutions without in any way first having established their social
desirability, or their environmental appropriateness.

I believe my own proposals steer a path between the extremes of
"Concordism" and the equally fruitless state where the general
public is involved in highly technical consultations of which it can
have only the faintest grasp. As I propose it, the public is to be very
directly involved in the regular policing of DED-free areas and in the
nationwide survey of where exactly wild elms currently grow best.
The national reserves, however, would be planned and directed by
experienced arboriculturalists.

I am, therefore, essentially proposing a genuine cooperation
between public and expert. My overriding aim is to create, while
there is still time, the only circumstances in which we may both understand what it is we have to lose, and generate measures by which to
oppose that loss.

The amount of leisure at the disposal of the average European and
North American individual is always increasing. It has been estimated that in forty years' time only a tenth of the West's present
labor force should be required to keep industrial production up to
present levels. A three-day working week will probably be a commonplace feature of western society just a few years hence. Already, there
is no lack of volunteer conscripts to conservation task forces of one
kind or another. But there are many would-be helpers who fail to
find an organization which operates on a scale they feel they can
believe in or one which credits them with the ability to contribute
meaningfully to conservation planning, rather than simply giving
them odd jobs to do in their spare time.

The technical know-how required to participate in most of the activities of an elm reserve of the kind I have envisaged could be picked up by an intelligent layman in a weekend of courses and demonstrations. In other words, the cost of an elm-reserve system need not, as far as labor costs are concerned, be estimated at commercial rates, since the need for permanent professional staffs could be rather small.

If, however, I am quite wrong in supposing that the manpower needed to create and run reserves could consist mainly of volunteer workers, I still consider that the cost of national elm reserves could be readily met — that is, even if they were manned entirely by professional woodsmen.

Direct taxation is only one among many potential sources of finance. But it is worth pointing out that the creation of one elm reserve of the kind portrayed above need only, for instance, cost taxpayers less than one cent a year, if $20 per tree per year is regarded as a realistic estimate of the running costs of a professionally managed reserve. Individual contributions would be miniscule.

An appropriate additional source of finance ought, I suggest, to be the timber industry itself. It is, after all, largely on account of the ease with which cheap foreign timber can be imported into developed countries that native hardwood trees like the elm do not automatically earn themselves careful protection as an important domestic source of timber.

At least part of the revenue from, say, a system of timber import licenses ought to be earmarked to conserve stocks of native trees, including normally unmarketed trees. It is high time that this industry put more capital into basic research and development aimed at conserving and marketing native hardwoods. At present such capital goes into foreign hardwood market research and at home into the expansion of commercial softwood-growing enterprises toward unattainable optima.

Ways should also be sought in which funds already spent on DTD-research or parochial DED-control schemes could be more usefully allocated, though not by *force majeure*. Municipalities and amenity groups which already operate successful disease-suppression programs or laboratories committed to DED-research programs have every right to carry on with them. But people who finance or operate schemes designed to preserve elm trees in perpetuity (that is, as permanent features of our landscapes) could ask themselves whether their efforts and resources might — in the long run — be more effectively applied as a part of a broadly based national strategy.

Charitable appeals, modeled on the "adopt-a-tree" theme, which some towns have successfully promoted to help pay for the upkeep of amenity trees, could be launched on a national scale, using the same theme in respect of trees whose amenity value was not immediately obvious. This is one way in which people unable to join directly in practical attempts to establish national elm-conservation areas could make their contribution.

Large industrial corporations and unions also have an important role to play — not just as sources of funds but more importantly as guardians of a vigilant approach to the protection of native wildlife, such as the elm, from imported diseases. College courses in economics and social sciences are now a normal part of the training of aspiring business managers and union officials. Surely it is time that environmental studies were added to their training programs. How else will we evolve a sensible policy toward potential environmental risks such as the importation and transport of insanitary timber? No DED-control program, whatever the scale of its operations, can be completely effective unless it is quite secure from foreign strains of DED fungus or imported species of DED-carrying bark beetles.

There may be a less-than-bright future for the suggestions I have made here and for pleas other commentators have made for a new approach to DED control. But there is nothing to prevent caring individuals making their own stand against the fashionable defeatist view that the elm's only hope for tomorrow lies nowhere but in a few cities and a few research centers. At Kew, London, one of Europe's foremost botanical gardens, a large stock of young elms of mixed pedigree is constantly maintained in anticipation of a solution to DED. There is no reason at all why any farmer or elm owner anywhere should not follow this scholarly example and take active steps to keep aside a territory for elms alone, disease or no disease. In their more enigmatic fashion, organizations, such as the Society for Celebrating the Year 2000, have for some time been transplanting elm seedlings grown in backyards and window boxes into overlooked corners of agricultural fields and parklands. I find such idealism has more to be said for it than the spurious inventiveness of spokesmen for "new landscapes" who seek to ignore the elm's existence altogether.

As tropical Third World countries continue to become industrialized, they require more and more of their own forest products for domestic consumption. Amounts of hardwood timber available for export are therefore bound gradually to dwindle, until the point is reached where it again becomes economic for developed countries to invest in native sources of supply. The greater the numbers and diversity of hardwood trees at large in the countryside at that time, the less expensive such ultimate investment must surely be.

The softwood timber market, barometer of the world timber trade, has problems of its own. For example, Russia recently doubled the cost of her export timber to alleviate the economic side effects of poor grain harvests.

As imported timber becomes ever more expensive, a great deal of effort is being made by all western countries to become as self-sufficient in forest products as possible. The method is that of injecting funds into the creation and expansion of large-scale commercial softwood forestry enterprises. Hardwood forestry — and therefore, of course, elmwood forestry — enjoys little spinoff from this rush of development. The viability of small-scale silviculture or agro-silviculture, whatever the nature of its products, continues to lag further and further behind that of big forestry.

Yet still, despite the almost universal intensification of commercial forestry in developed countries, all countries show a continual net loss of tree cover every year. This loss reflects above all, I suggest, the effects of disease, fire, neglect and urban development on small noncommercial woodlands — consisting largely of native hardwoods like the elm.

Direct comparisons between the plight of the threatened elm and the plight of huge tropical rain forests are not so farfetched as they might seem at first thought. In both cases the sweeping assumption is made that particular trees are unlikely to be worth more to us alive than dead.

The catch is plain. Until large-scale conservation experiments are made, little can be done to make a clear-cut case for conservation as a matter of course. Yet destruction (whether deliberate or, as in the case of the elm, negligent) proceeds at such a rate that the study of its undesirable consequences becomes daily more academic and more pointless.

We can, if we wish, still save the elm from DED. We can still save other native trees from similar disease problems. But first we must liberate ourselves from our tendency to view trees merely as ornaments, or as more-or-less convenient sources of wood. This is a view which does justice neither to the environmental sovereignty nor to the usefulness of trees, let alone to their beauty.

When each of us thinks of what we stand to lose by deforestation, we are bound to think both big and small. Of the woodland sights and sounds imprinted on our senses during many a childhood vacation or picnic. Of the dark immensity of the many-tiered rain and cloud forests of the tropics. Of the crisp feel of wood fiber in the opening page of a new book. Of the never-still edge of the world's deserts as they encroach on fertile land, unchecked by the ranks of tough, dry-country trees and shrubs which used to hold them back. Of the swish of the carpenter's plane across the fresh surface of seasoned timber — man's oldest and most versatile construction material.

As matters stand now, different thoughts and feelings cannot be harmonized within a single point of view, a sane consensus from which a practical solution to the world problem of deforestation can arise. The interests which govern the protection and the exploitation of trees pull in too many different directions. The reconciling of these interests so as to secure a proper future for trees and — it follows — for ourselves, is the great challenge of our century.

The confused and universal danger of deforestation . . . our individual and alive response to it: which will win?

Footnotes

1 These last amounts include the respiration and decay products from natural animal sources. Animals are, however, in a sense plant products and their small-scale contribution is anyway not worth treating separately in this context.

2 Walter Nichol, a Scottish pundit of landscape design, deplored in *The Planter's Kalender*(21) of 1810-12 the use of English elm — "that ugly and disgusting hedge-timber" — in planned vistas. "They never make the best trees . . . they always produce suckers from their roots and disfigure the ground on which they stand."

Nichol recommended that the English elm "ought to be budded and grafted on the Scots elm [i.e., the fuller-figured and nonsuckering Wych elm]: in this way trees of superior vigour and figure would be obtained which would never produce a sucker. . . ."

The strictures of Nichol and others were taken to heart by many landowners. Hence the confusion which now besets modern efforts to classify British elms — their pedigree is often lost in a long past of hybrid experimentation.

The English or red elm produces quantities of seeds but, unlike those of the Wych elm, they are not viable in their native soil. Suckering is almost the only reproductive resource the tree has. In warmer climates, curiously enough, the seeds are apparently viable: Helen Bancroft(1) has described English elms grown from seed in the royal gardens of Spain. The infertility of English elms in England has been advanced as an argument for their having been introduced from elsewhere — some say by the Romans. What is more likely is that the English elm is endemic but the English climate has changed for the colder within historic time. Vines and other warm-climate crop plants used to thrive in southern England, where now the climate can support only hardy modern hybrid stock. It has furthermore been shown that for some plants (e.g., certain thistles) a drop in mean summer temperatures of less than one degree centigrade can mean sterility for the seeds.

3 At the time of going to press the decline of rook populations and the relationship of that decline to the DED epidemic has, in fact, been confirmed(28).

4 The most telling example of juvenility is the region of juvenile leaf growth in beech trees, which extends about three meters up from the base of the tree and is most evident in the brown leaves beeches retain in winter long after the rest of their leaves have fallen.

Three meters is also the approximate height of the frost line, where ground frost ends and air frost takes effect in extremely cold weather. S.A. Searle(31) maintains that beeches possess genes that equip them to grow during an ice age, not as a forest tree, one to two hundred feet tall, but as a dwarfish shrub no taller than about three meters. A stand of beeches planted out in areas of very testing climate will, in fact, nearly always take this form, in preference to the form more familiar to us.

Juvenile growth in other trees, such as holly and European ivy (a creeper well able, under certain circumstances, to grow as a tree), manifests itself in leaves the very shape of which is different from that of leaves above the three-meter line on the same tree. So marked is this tendency in the ivy that it was long thought to be not one, but two separate species.

The functional reason for this capacity to arrest growth and change identity seems to be to prevent the tree from flowering in adverse conditions when pollination is out of the question: only when a beech has grown above three meters do its flowers begin to appear and become sexually viable. If trees like beeches, with their averagely slow succession of generations, are imprinted with the relics of such a mechanism, it seems reasonable to suppose that all trees which have confronted ice age conditions during their evolution may be able to make similar adjustments to some extent or another.

Appendix

Organizations Currently Involved With Tree Conservation

Acres, Inc.
1802 Chapman Rd.
Huntertown, IN 46748
(219) 637-6264

Agricultural Research Institute
2100 Pennsylvania Ave., N.W.
Washington D.C.
20037

Alaska Conservation Soceity
Box 80192
College, AK
99701

**American Association for
    Conservation Information**
c/o Ronald E. Shay
Oregon Game Commission
Portland, OR
97208

**American Committee for
    International Conservation**
c/o The Wildlife Society
3900 Wisconsin Ave., N.W., Ste. S-
    176
Washington, D.C.
20016

American Forest Institute,
1619 Massachusetts Ave., N.W.
Washington, D.C.
20036

American Forestry Assoc.
1319 18th St., N.W.
Washington, D.C.
20036

**America's Future Trees
    Foundation**
1200 Hanna Building
Cleveland, OH
44115

**Association of Conservation
    Engineers**
c/o Ronald Hansen
Missourie Dept. of Conservation
P.O. Box 180
Jefferson City, MO
65101

**Association of Consulting
    Foresters**
Box 6
Wake, VA
23176

**Association for the Protection of
    the Adirondacks**
21 E. 40th St.
Room 704
N.Y., N.Y.
10016

Big Thicket Assoc.
Box 198
Saratoga, TX
77585

**Citizens Committee on Natural
    Resources**
1346 Connecticut Ave., N.W.
Washington D.C.
20036

Citizens for Conservation
1134 S. Washington Ave.
Lansing, MI
48910

**Conservation Education
    Association**
c/o Dr. Robert Cook
Office of Dean of Colleges
Univ of Wisconsin — Green Bay
Green Bay, WI
54302

Conservation Foundation
1717 Massachusetts Ave., N.W.
Washington, D.C.
20036

Conservation League
110 W. 71st St.
N.Y., N.Y.
10023

**Conservation & Research
    Foundation**
Box 1445
Connecticut College
New London, CN
06320

Conservation Services
S. Great Road
Lincoln, MA
01773

Elm Research Institute
Harrisville, N.H.
03450

**Federation of Western Outdoor
    Clubs**
4534½ University Way, N.E.
Seattle, WA
98105

**Forest Farmers Association
    Cooperative**
Four Executive Park E., N.W.,
    Rm. 380
Atlanta, GA
30329

Forest History Society
P.O. Box 1581
Santa Cruz, CA
95061

**Forestry Conservation
    Communications
    Association**
c/o R.E. Walker
Florida Div. of Forestry
Collins Building
Tallahassee, FL
32304

Friends of the Earth
529 Commercial St.
San Francisco, CA
94111

Friends of Nature
Brooksville, ME
04617

Friends of the Wilderness
3515 E. Fourth St.
Duluth, MN
55804

Greensward Foundation
104 Prospect Park, W.
Brooklyn, N.Y.
11215

**International Union for Conserva-
    tion of Nature & Natural
    Resources**
1110 Morges,
Switzerland

Izaak Walton League of America
1800 N. Kent St.
Suite 806
Arlington, VA
22209

J.N. "Ding" Darling Foundation
610 Fleming Bldg.
Des Moines, IA
50309

**National Association of Conserva-
    tion Districts**
1025 Vermont Ave., N.W.
Washington D.C.
20005

**National Association of State For-
    esters**
Dir., Office of Forest Resources
Dept. of Natural and Economic
    Resources
P.O. Box 27687
Raleigh, N.C.
27611

National Audubon Society
950 Third Ave.
N.Y., N.Y.
10022

**National Bark Producers Associa-
    tion**
1750 Old Meadow Rd.
McLean, VA
22101

National Wildlife Federation
1412 16th St., N.W.
Washington, D.C.
20036

**Natural Resources Council of
    America**
1025 Connecticut Ave., N.W.,
    Suite 911
Washington, D.C.
20036

Nature Conservancy
1800 N. Kent St., Suite 800
Arlington, VA
22209

New Roots for Young America
c/o George C. Whitford
Reliance Insurance Co.
4 Penn Center Plaza
Philadelphia, PA
19103

North American Habitat Preservation Society
P.O. Box 869
Adelphi, MD
20783

Ozark Society
P.O. Box 2914
Little Rock, AR
72203

Ranger Rick's Nature Club
1412 16th St., N.W.
Washington, D.C.
20036

Resources for the Future
1755 Massachusetts Ave., N.W.
Washington D.C.
20036

Save-the-Redwoods League
114 Sansome St.
Room 605
San Francisco, CA
94104

Society of American Foresters
1010 16th St., N.W.
Washington D.C.
20036

Soil Conservation Society of America
7515 N.E. Ankeny Rd.
Ankeny, IA
50021

Southern Forest Institute
One Corporate Sq., N.W.
Suite 280
Atlanta, GA
30329

Student Conservation Association
Olympic View Dr.
Rte. 1, Box 573-A
Vashon, WA
98070

U.S. Environment and Resources Council
12412 Shelter Ln.
Bowie, MD
20715

Western Forestry & Conservation Assoc.
1326 American Bank Building
Portland, OR
97205

Wilderness Society
1901 Pennsylvania Ave., N.W.
Washington, D.C.
20006

For practical information about treatment of DED, write to:
Forest Service, U.S. Dept. of Agriculture, 6816 Market St., Upper Darby, P.A. 19082, or:
Elm Research Institute, Harrisville, N.H. 03450, U.S.A., or, in the U.K., to:
Forestry Commission Research Station, Wrecclesham, Farnham, Surrey.

Sources used for the Tables and Figures are:
Table 1 *(p. 16)* Simms (32).
Table 2 *(p. 17)* Curry Lindahl (3) and UN sources.
Figure 1 *(p. 24)* D.H. and M.P. Tarling *Continental Drift*, Bell 1971.
Figure 3 *(p. 50)* Lemon *et. al., Bioscience* 20 (19), 1970.
Figure 7 *(p. 97)* G.W. Dimbleby, 1969; in *Science in Archaeology*, 169.
Table 4 *(p. 126) Pulp and Paper International*, Review Number 1977.
Table 5 *(p. 148)* Douglas and Hart (4), and other standard sources.
Figure 10 *(p. 154)* Thèodore Monod, *Mitteilungen als der Botanischen Staatssammlung. Munchen* 10: 375-423, 1971. (After C. Troll.)
Figure 11 *(p. 194)* based on studies made by W.N. Cannon and D.P. Worley (1976) in thirty-nine US municipalities of various sizes.

Bibliography

1. BANCROFT, Helen, "The Elm Problem," *Quarterly Journal of Forestry*, April 1935: 1-5.
2. CARSON, Rachel, *Silent Spring*, New York, Houghton Mifflin Co., 1962.
3. CURRY-LINDAHL, Kai, *Conservation for Survival. An Ecological Strategy*. New York, William Morrow & Co., 1972.
4. DOUGLAS, J. Sholto & HART, A. de J., *Forest Farming*, Emmaus, Pa., Rodale Press Inc., 1978.
5. FRAZER-DARLING, Sir Frank, *Wilderness and Plenty — The Reith Lectures 1969*, London, Oxford University Press, 1970.
6. GOODLAND, R. & BOOKMAN, J., "Can Amazonia survive its highways?" *Ecologist* 7: 376-80, 1977.
7. GRIBBEN, J., *Forecasts, Famines and Freezes. Climate and Man's Future*, New York, Walker & Co., 1976.
8. HAFFER, J., "Speciation in Amazonian forest birds", *Science* 165, 3889:131-37, 1969.
9. HEYBROEK, H., p. 68 *in* "Dutch Elm Disease". *Proceedings of IUFRO Conference, Minneapolis — St. Paul, September, 1973*, USDA Forest Service, 1975.
10. HOOPER, M.D., "The Botanical Importance of our Hedgerows", pp. 58-61 *in* Perring, F. (ed.) 1970 *see below*.
11. HUXLEY, Aldous, *The Olive Tree and Other Essays*, London, Chatto and Windus, 1936.
12. HUXLEY, Anthony, *Plant and Planet*, London, Allen Lane, 1974.
13. JEFFERIES, Richard, *Wild Life in a Southern County*, London, Edinburgh, Dublin, New York, Thomas Nelson 1879.
14. LEWIN, R. "Why the Yule Logs must not burn", *New Scientist* 23-30 December, 1976: 750-751.
15. MARQUARDT, G., "Die Schleswig-Holsteinische Knicklandschaft", Sch. Geogr. Inst. Univ. Kiel, 13(3): 1-90, 1950.
16. McCORMICK, J., *The Living Forest*, New York, Harper and Row in collaboration with the American Museum of Natural History, 1959.
17. McCREA, W.H., "Ice Ages and the Galaxy", *Nature* 255: 607, 1975.
18. McNAUGHTON, S.J., "Stability and diversity of ecological communities", *Nature* 274, July 20, 1978: 251-252.
19. MAY, R.M., *Stability and Complexity in Model Ecosystems*, Princeton University Press, 1973.
20. MENSCHING, H., "Drought disaster and desertification and their impact on the ecosystem of the Sahelian zone between Air and Darfur", *Proceedings of 3rd International Conference on the Central Bilad Al-Sudan*, Khartoum University, November 1977.
21. NICHOL, Walter, *The Planter's Kalendar*, Edinburgh, Willison, 1812.
22. PERRING, F. (ed.), *The Flora of a changing Britain*, BSBI Conference Report No. 11, Botanical Society of the British Isles, 1970.
23. PORTERES, R., "Formations prairiales paléotropicales à *Themeda* et

leur extension ancienne en Afrique", *Compte rendu sommaire des Séances de la Société Biogéographique de France* 109: 241-243, Paris, 1951.

24. RICE, Colin, "Blind alleys and open prospects in forest economics", *Forestry*, 49: 99-107, 1973.

25. RICHARDS, Paul W., "Doomsday for the world's tropical rain forests?" *Unesco Courier*, October, 1975: 16-3

26. ROBINSON, Gordon, "Forestry as if trees mattered: a bold stand", *Not Man Alone* (Journal of US Friends of the Earth), August 1976, pp. 24ff.

27. ROLSTON, L.H. & McCOY, C.E., *Introduction to Applied Entomology*, New York, Ronald Press Co., 1966.

28. SAGE, B. "The Rook in Britain", *New Scientist*, June 29, 1978: 898-899.

29. SCHUMACHER, E.F., *Small is Beautiful*, Blond & Briggs, 1973.

30. SCHWEITZER, E.M., "Comparative Anatomy of Ulmaceae" *Journal of the Arnold Arboretum* 52: 524-573, 1971.

31. SEARLE, S.A., *Environment and Plant Life*, London, Faber & Faber, 1973.

32. SIMMS, Eric, *Woodland Birds*, London & Glasgow, Collins, 1971.

33. SLOANE, Sir Hans, *A voyage to the islands Madera, Barbados, Nieves, S. Christophers and Jamaica* ... London 1707-25, Royal Society.

34. THOMAS, Lewis, *The Lives of Cells*, New York, Viking Press Inc., 1974.

35. TURNBULL, Colin, *The Forest People*, London, Chatto & Windus, 1961.

36. UNESCO, Program on Man and the Biosphere, International Working Group on Project 1: *Ecological effects of increasing human activities on tropical and sub-tropical forest ecosystems*, Rio de Janeiro, February 11-15, 1974, Final Report (16) UNESCO, Paris.

37. WHITMORE, T.C., *Tropical Rain Forests of the Far East*, New York, Oxford University Press, 1975.

38. WILKINSON, G., *Epitaph for the Elm*, Salem, New Hampshire, Hutchinson, 1978.

Index